Principles of Sedimentology and Stratigraphy

Principles of Sedimentology and Stratigraphy

Cody Long

STATES
ACADEMIC PRESS
www.statesacademicpress.com

States Academic Press,
109 South 5th Street,
Brooklyn, NY 11249, USA

Visit us on the World Wide Web at:
www.statesacademicpress.com

ISBN: 978-1-63989-441-3 (Hardback)

Cataloging-in-Publication Data

Principles of sedimentology and stratigraphy / Cody Long.
p. cm.
Includes bibliographical references and index.
ISBN 978-1-63989-441-3
1. Sedimentology. 2. Geology, Stratigraphic. 3. Sediments (Geology). 4. Physical geology.
5. Sedimentary structures. I. Long, Cody.

QE471 .S43 2022
552.5--dc23

Table of Contents

Preface

The study of sediments such as sand, silt and clay is referred to as sedimentology. This includes the analyzes of various processes related to their formation, transportation, diagenesis and deposition. Sedimentary rocks and structures are the key areas of focus under this discipline. Sedimentary rocks can be broadly classified into carbonates, clastic rocks, chemical sedimentary rocks and evaporites. Stratigraphy is a branch of geology which is closely related to sedimentology. It focuses on the structure of rock layers and layering. This discipline can be divided into lithostratigraphy, biostratigraphy and chronostratigraphy. The topics included in this book on sedimentology and stratigraphy are of utmost significance and bound to provide incredible insights to readers. Some of the diverse topics covered herein address the varied branches that fall under this category. Coherent flow of topics, student-friendly language and extensive use of examples make this book an invaluable source of knowledge.

A detailed account of the significant topics covered in this book is provided below:

Chapter 1- The branch of study which focuses on sedimentary rocks as well as the processes through which they are formed is called sedimentology. It also studies sediments such as sand, silt and clay. The diverse aspects of sedimentology as well as sedimentary basin and sedimentary petrology have been introduced briefly in this chapter.

Chapter 2- The naturally occurring materials which are broken down through the processes such as erosion and weathering are known as sediments. Some of the different types of sediments are biogenic sediments and marine sediments. The topics elaborated in this chapter will help in gaining a better perspective about these types of sediments as well as sedimentary rocks.

Chapter 3- Numerous processes are studied within the discipline of sedimentology. A few of them are sediment transport, deposition, erosion, diagenesis and weathering. The movement of organic and inorganic materials through the action of water or gravity is referred to as sediment transport. All these diverse sedimentology processes have been carefully analyzed in this chapter.

Chapter 4- The features of sediments and sedimentary rocks, which are formed at the time of deposition, are called sedimentary structures. They can be broadly categorized into primary sedimentary structures and secondary sedimentary structures. This chapter has been carefully written to provide an easy understanding of these types of sedimentary structures.

Chapter 5- The branch of geology which deals with the layers of rocks and layering is called stratigraphy. A few of the major concepts within stratigraphy are successions of strata, stratigraphic nomenclature and Steno's principles of stratigraphy. All the diverse concepts of stratigraphy have been carefully analyzed in this chapter.

Chapter 6- Sedimentary rocks such as limestone and sandstone have a diverse range of applications in a wide variety of industries. A few of these are dimension stone, ceramics, petroleum geology and aquifers. This chapter closely examines these key applications of sedimentary rocks to provide an extensive understanding of the subject.

It gives me an immense pleasure to thank our entire team for their efforts. Finally in the end, I would like to thank my family and colleagues who have been a great source of inspiration and support.

Cody Long

Chapter 1

Sedimentology: An Introduction

The branch of study which focuses on sedimentary rocks as well as the processes through which they are formed is called sedimentology. It also studies sediments such as sand, silt and clay. The diverse aspects of sedimentology as well as sedimentary basin and sedimentary petrology have been introduced briefly in this chapter.

Sedimentology is the scientific study of sedimentary rocks and the processes by which they were formed. It is concerned with the composition and genesis of sediments and sedimentary rocks, and the creation of predictive models that resemble and predict sedimentation. It's used to describe sedimentation processes by interpreting the facies in order to know the characteristics, methodology and principles of sedimentation. In addition, sedimentology is related to straigraphy which is the study of the physical relationships of sedimentary layers. The study of sedimentoloy and stratigraphy together gives clearer vision about the layers and provides us with any information we need.

Sedimentary Processes

Sedimentary processes lead to the development of sediments and they consist of 3 main processes:

- Physical processes (weathering and erosion).
- Chemical processes (precipitation of chemical compounds in water solutions).
- Biological processes.

These 3 processes are important parts of the development of layers. Weathering and erosion are related to sedimentation processes through the material they provide. This material is transported to deposition place and imbedded there. Diagenesis processes transform sediments into sedimentary rocks due to the physical and chemical changes. The accumulation of these sedimentary rocks is subjected to temperature and pressure which take place in the chemical processes affecting the rocks. Then, they are subjected to biological processes which affect the growth of the rocks.

Sedimentary Environments and Facies

Facies of sedimentary rocks means that every rock has special specifications in which we can know the history by which it's formed. Consequently, engineers and geologists can use the history of these rocks in correlating and in predicting the existence of minerals. There are many factors that help in knowing the facies of rocks and the most three important of them are:

- Minerology.
- Grain Size and Sorting.
- Angularity.

There are three sections that geologists use in facies identification:

- Facies Analysis (Interpreting the layers of rock in the ground).
- Facies Association (Establishing several facies according to their combinations).
- Facies Successions (Facies associations with unique characteristics).

Spectrum of Environments and Facies

The discription of sedimentary rocks through their facies and environments results in grouping every similar rock to be unique and different from others, and hence, we can categorize the sedimentary rocks to make the process easier.

Classification of Sediments and Sedimentary Rocks

Sedimentary rocks can be classified into two main groups based on the processes by which they were derived Clastic and Non Clastic rocks.

Clastic Rocks (>80%)

They are particles derived from the weathering and erosion of precursor rocks and consist primarily of fragmental material and are classified by grain size and composition. Clastic rocks are the most common (i.e. mudstones (65%) and sandstones (20-25%)). They are classified into:

- Volcaniclastic.
- Terrigenous clastic.
- Carbonates.

Non Clastic Rocks

Non clastic represents about 10 - 15% of sedimentary rocks.They are classified into:

- Carbonates (in common with clastic rocks).
- Others (i.e. coal).

- Evaporites (chemical depositions).

Sedimentation Principles, Methodology and Characteristics

Principles of Sedimentation

Principles of sediments are used to know more information about the characteristics of sedimentary rocks that are deposited in basins. Principles of Sedimentary rocks can be classified into:

- Principle of superposition: Sedimentary layers are deposited in a time sequence, with the oldest on the bottom and the youngest on the top.
- Principle of original horizontality: The deposition of sediments in a horizontal way which is their relief angle.
- Principle of lateral continuity: The layers of sediments extend in all directions If there is no any restrictions on their ways.
- Principle of cross-cutting relationships: Any intrusive/cutting rock or layer is younger than the layers of sediments.

Methodology of Sedimentation

Methodology of identifying the sedimentation of rocks is based on collecting all possible information from the rocks including:

- Measuring and describing the outcrop and distribution of the rock unit.
- Describing rock core.
- Identifying sequence stratigraphy characteristics (Describes the progression of rock units within a basin).
- Describing the lithology of the rock.

Characteristics of Sedimentation

Characteristics of sedimentary rocks are very useful for identifying sedimentation methodology as it is possible to get more information about rocks. Examples of the characteristics include:

- Porosity is the volume of voids within a rock which can contains fluids.
- Permeability is the ability of water or other liquids to pass freely through a rock.
- Roundness indicates the roughness of sedimentary grains' surfaces.
- Sorting indicates to the range of particle sizes in the sediments or sedimentary rock.

- Matrix is a fine-grained material that's deposited originally with coarser-grained minerals.

Sediment Grain Size and Analysis

Grain size represents a fundamental character of clastic sedimentary rock and one of the most important properties for description of these rocks. Sizes of clastic grains reflect their weathering and erosional history, besides providing idea about their transportation motif i.e. as bed load, saltation or suspension. By the term grain size normally we mean grain diameter. This is true when the grain is perfectly spherical. However, most grains in nature are non-spherical and that generates most relevant question 'what is the representative diameter for a non-spherical grain?' To resolve this question the concept of 'nominal diameter' is proposed. The 'nominal diameter' of a grain is the diameter of a sphere that has equal volume as that of the grain. Again, in practice we are seldom in a position to measure volume of each grain, and therefore use indirect method for measurement of any grain. If a grain resembles tri-axial ellipsoid, we measure its long axis (a), intermediate axis (b) and short axis (c) for further use of data.

As in nature grain sizes of sediments vary widely i.e. from clay size less than 0.002mm in diameter to boulder size more than 25cm in diameter, necessity was felt to categorize grains in terms of size scale.

Grain size Scales

1. Udden-Wentworth Scale: Proposed by Udden in 1914, the scale divides consecutive classes of grain sizes in millimeter scale by multiples of 2 on higher side and decrease by multiple of 1/2 on the lower side.

Boulder I	Cobble I	Pebble I	Granule I	Sand I	Silt I	Clay
256	64	4	2	1/16	1/256	(mm)

As commonly available grain sizes in nature falls in fine-size category i.e sand to clay and their class intervals comes in fraction following this scale, computations became very cumbersome in early days in absence of any high speed computational facility. This gave birth to another grain size scale which is referred as Phi scale.

2. Phi (Krumbein) Scale: Krumbein in 1934 devised this scale that is referred as phi (Φ) scale.

Phi (Φ) = $-\log_2 d$; where d is diameter of a grain in mm.

The use of 2 as the base of the logarithm allowed direct conversion of Udden's grade scale to Krumbein phi scale. Also, inversion of sign transformed the fractions to positive, round integers, thereby making computations much easier for the finer grains (sand, silt, clay).

Wentworth nomenclature	Boulder		Cobble		Pebble		Granule		Sand		Silt		Clay	
Udden's grade scale		256 (2^8)		64 (2^6)		4 (2^2)		2 (2^1)		1/16 (2^{-4})		1/256 (2^{-8})		mm
Krumbein's phi-scale		-8		-6		-2		-1		+4		+8		ϕ

Conversion of Udden-Wentworth scale to Krumbein phi scale.

Measurement of Grain Size

Gravel and Sand Size

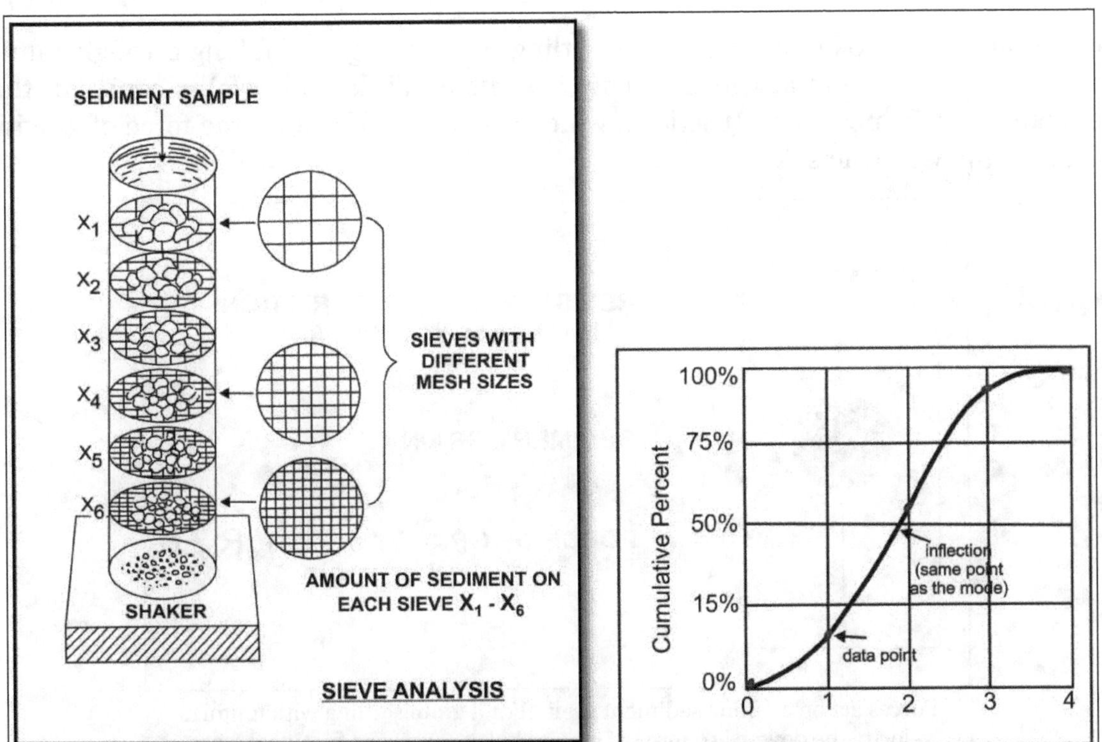

Bank of seives with decrease in opening size downward. Depending on number of seives used and their respective opening sizes, the original sediment samples get divided into a number of size fractions on seiving.

Measurement of sand and gravel grains are done by means of mechanical seiving. Seives are screens of standard size and specified by the American Society of Testing Materials (A.S.T.M). Openings of screens increase uniformly in multiples of $\sqrt[4]{2}$ following size grades of Wentworth size scale, starting from an opening of 0.0029 inch (200 mesh). For size measurement, a bank of seive is taken with downward decreasing opening (mesh) size. Sediment with measured weight is kept on the topmost seive and then the seive bank is mechanically or electrically shaken. Depending on the size of grains, the sediment get distributed among the seives used; grains larger than the seive size will remain on the seive and grains with smaller size pass on to the seives further down. By weighing fraction of sample lying on each seive, a grain size distribution curve is constructed. The lower limit of seive analysis is 0.04-0.03 mm, grains finer than this size show large amount of cohesion and do not allow measurement of individual grain by this method.

Silt and Clay Size Fraction

Sediments of this size fraction cannot be measured by seive analysis because of their high cohesive character. In lieu, measurement of size for these sediments are done through number of other methods, most of which deals with settling velocity of sediment grain in liquids following Stoke's law.

According to the Stoke's law when the settling velocity of grains falling through a uniform solution (solution at which density is same at all depth levels) is constant, the resistance to the movement (friction), which acts upward balance the force of gravity on the grain, which acts downward.

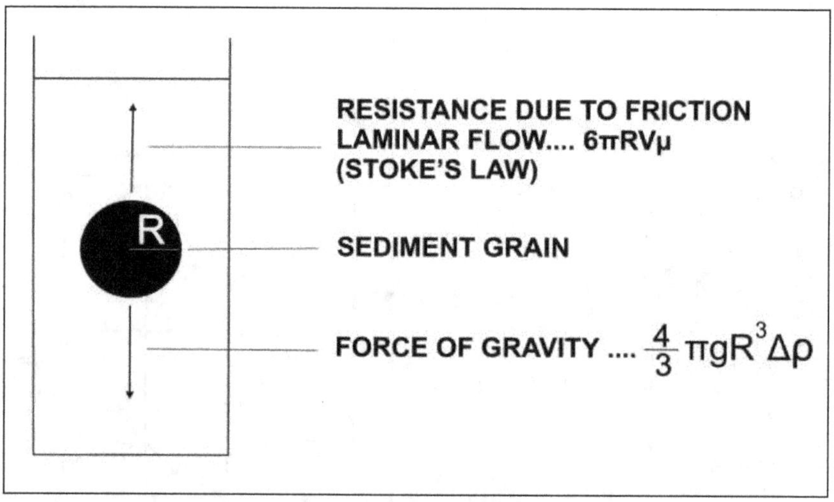

Forces act on a falling sediment grain. For a grain settling with uniform velocity the downward force of gravity balance upward frictional force.

$$\underset{\text{(Friction)}}{6\pi R v \mu} = \underset{\text{(Gravity)}}{4/3\pi g R^3 \Delta\rho}$$

$$v = 2/9\mu g R^2 \Delta\rho$$

Where μ is viscosity and Δρ is density difference between the grain and liquid. R is the radius (in cm) of sediment grain. From the relation it can be noticed that the settling velocity proportionally varies with the radius of the grain ($V \propto R^2$) as π, g and $\Delta\rho$ are constant between the grain and liquid.

For settling through water, the relation becomes $\log v = 2 \log R + C$

This relationship applies well for grains smaller than 0.1mm in diameter. With increase in grain size (i.e. 0.1 to 0.5mm) settling velocity increases and turbulence develops around the grain. Settling velocity of such relatively large size grains varies proportionally with square root radius of the grain.

However, measurement of settling velocity of each grain is also not a practicable idea and hence, attempts were made to overcome the difficulty in some indirect way by use of hydrometer or sedimentation balance. Other methods which are in use in recent times include refraction or dispersion of light through suspensions of sediment (scattering) or use of x-ray instead of light.

1. Hydrometer: A suspension of sample is created with a mixer in a cylinder so that at time T_0 there is an even distribution of grains in the solution and density of solution is same throughout the cylinder. A hydrometer will float in the solution and register density of the solution on a scale on the upper part of the tube. If the suspension is allowed to stand for a time T_1, grains will settle in the bottom of the cylinder.

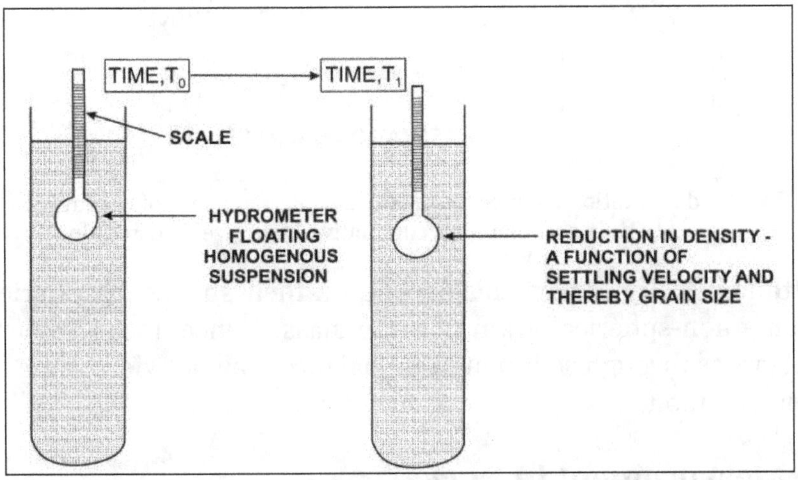

Operational procedure of a hydrometer. Note change in degree of submergence of hydrometer in a solution with time depending on change in density of solution.

At the level at which the bulb of the hydrometer is floating the largest of the grain sizes already must have settled resulting decrease in density of the solution. This reduction in density of solution gets registered by further submergence of hydrometer and recorded in the scale at the upper part of tube. By taking successive readings of the density we plot a density variation curve with respect to time. Since density reduction is a function of settling velocity, the density variation curve can be calibrated as a grain-size distribution curve. In case the sediment contains a large clay fraction, use of centrifuge may be a more desired option. The acceleration term g in the Stoke's law increased in the centrifuge, which can be calculated from the velocity of rotation and length of rotating arm.

2. Sedimentation Balance: In this method sediments suspended in a solution in a cylinder fall through the water column and settle on a balance pan kept at any depth level within the cylinder. The balance pan is kept connected with a weigh meter. With passing time more and more sediment will get settled on the pan and the increasing weight of the pan get recorded in the weigh meter. The increase in weight is plotted against time, which indicated suspension settlement from solution as a function of time. This gives a direct cumulative curve that can be calibrated to give a grain-size distribution curve.

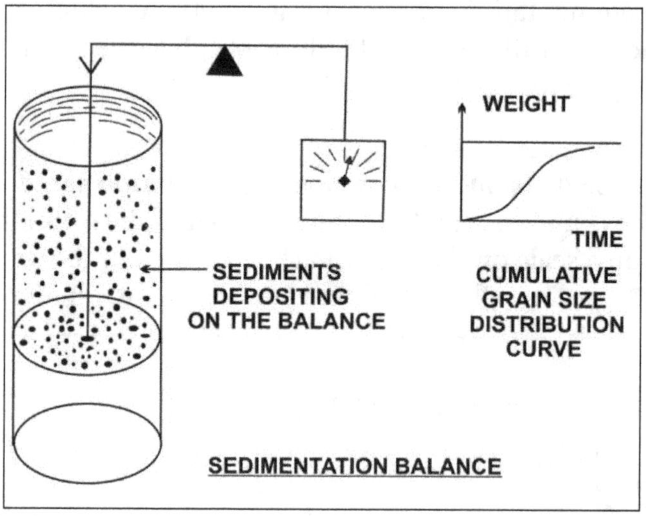

Use of sedimentation balance for measurement of increase of weight with respect to time and therefore cumulative grain size distribution.

However, settling velocity of particles depend on their shapes. A spherical grain settles faster than a non-spherical grain of same mass. Hence, measurement of size for non-spherical grains through sedimentation balance may not yield same result as that from the seiving method.

Grain Size Measurement in Rocks

Grain size measurement in sedimentary rocks is done in thin section by use of microscope. However, it is difficult to measure fine fractions (fine silt and clay) in this method. One major caution in this method is 'sectional effect' i.e. under thin section diameter of a grain will depend on relative orientation of grain and the cut section on which thin section is made and hence, in most cases we do not get the greatest diameter. With spherical grains, statistically expressed relation between real diameter d_r and observed diameter d_o is $d_r = 4/\pi d_o$. However, in practice we do not give this correction to measured grain sizes from a rock as sectional effect on all grains of a rock makes no change on their relative sizes.

Different Methods of Statistical Analyses of Grain Size Data

Histogram

A histogram is essentially a bar graph in which the percentages for each grade size are plotted as a column. It is very easy to prepare and one can easily interpret general features of the sediments. It is a pictorial method.

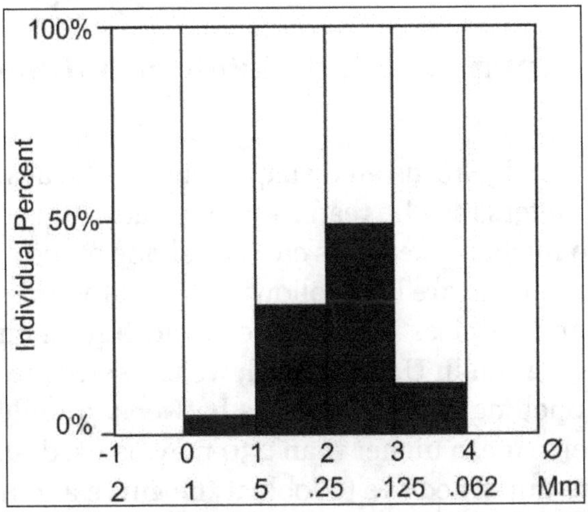

Sediments of different size grades plotted as columns in a histogram.

However, this method cannot be used for determination of any statistical parameters such as median, sorting etc. Its shape is greatly affected by sieve interval chosen. The same sample may look entirely different if it is analyzed on a different set of screens. Nevertheless, it proves of value plotting distribution of sediments on a map or stratigraphic section, as the height of the column may be more easily compared even by untrained eye.

Cumulative Curve, Arithmetic Ordinates

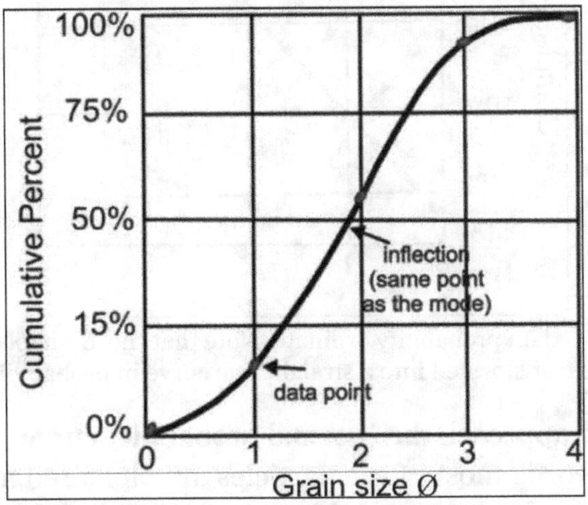

A cumulative frequency curve plotted with arithmetic ordinate.

This is the most commonly used method. In the abscissa, one may either millimeter (in which case he/she must use semi log paper) or phi units (ordinary squared arithmetic paper). In the ordinate arithmetic scale runs from 0 to 100%; grain size is plotted on the abscissa with coarser particles to the left (this is customary in all size analysis plots). Cumulative percentage of the sediments are plotted on this graph; for example, if 30% of the material is coarser than 2Φ caught on 2Φ screen) then 30 is plotted as the ordinates against 2.0 as abscissa. A curve is drawn through all the resulting points.

Curve Must Pass through all Plotted Points and Never use a French Curve

The sample analysis normally forms an S shaped curve. The advantage of this curve is that all statistical parameters may be read from it exactly, thus one can compare samples quantitatively as to median, skewness etc. The shape of the curve is independent of sieves used. The seive numbers are kept following U.S standard seives that correspond to various millimeters and phi sizes. Whereas seive mesh 10 (2 mm opening) and below correspond to gravel size grain in U.S standard, we use seive mesh between 10 (2 mm) and 230 (0.0625 mm opening) for sand grains between granule and fine sand grade. For coarse silt grains seive mesh higher than 230 may be used. Its only disadvantage is that it is difficult for the untrained eye to look at the curve and interpret it at a glance. Also if the sieve interval is wide, sketching the curve between data points is subject to considerable error.

Cumulative Curve and Probability Ordinate

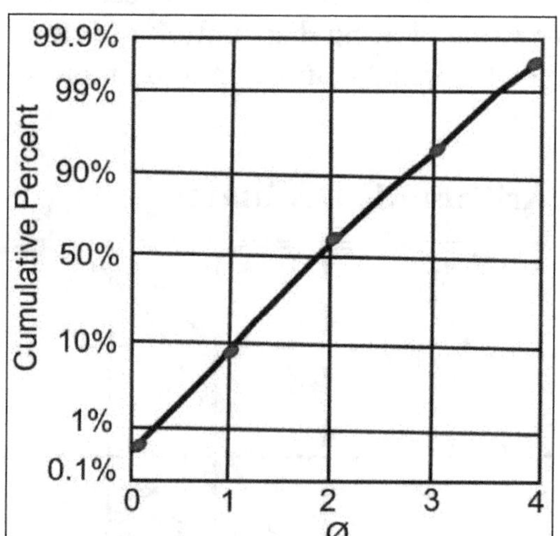

Cumulative curve plotted on probability ordinate. Note that the 'S' shaped curve in arithmetic ordinate has transformed into a straight line curve in probability ordinate.

Most sediments tend to approach the "normal probability curve" in their size frequency distribution, in other words most of the particles are clustered about a given size with less and less material on each side of this size. If the cumulative curve of sediment following the normal, symmetrical probability distribution is plotted on probability paper, the result is a perfectly straight line whose position depends on the average particle size and whose slope depends on the sorting. This happens because the probability scale is very condensed in the middle of the scale (30 to 70%) and very much expanded at the ends (under 10 or over 90 %), thereby straightening out the S-shaped curve which would result if arithmetic ordinates were used. Thus it is very valuable for studying the departure of sediments from the normal probability law. Moreover, since the "tails" are straightened out and the sample tends to plot as a straight line, it is possible to read the statistical parameters with much greater accuracy because of the ease of interpolation and extrapolation. Hence, this is the curve that must be used for all determination of parameters. The only disadvantage is that it is even less pictorial than the arithmetic cumulative curve.

Frequency Curve

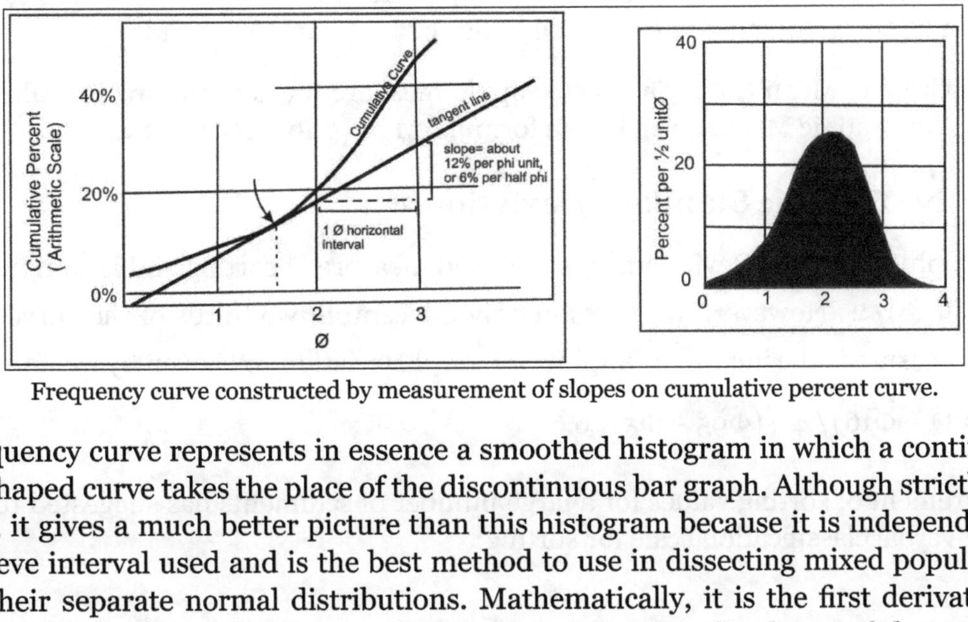

Frequency curve constructed by measurement of slopes on cumulative percent curve.

A frequency curve represents in essence a smoothed histogram in which a continuous bell-shaped curve takes the place of the discontinuous bar graph. Although strictly pictorial, it gives a much better picture than this histogram because it is independent of the sieve interval used and is the best method to use in dissecting mixed populations into their separate normal distributions. Mathematically, it is the first derivative of the cumulative curve, and is thus obtained by measuring the slopes of the tangents to the cumulative curve. To construct it, one plots a cumulative curve with arithmetic (not probability) ordinates. Now one measures the slopes of tangents to this curve at various grain-size values. For example, if one wants to find the frequency at a diameter of 2.5Φ, lay a straight-edge tangent to the curve at the point where the 2.5 Φ line intersects it. Measure the slope of this tangent by noting how much the tangent rises over a horizontal distance of 1/2 phi unit. This value then is plotted at 2.5 on the frequency curve. It must be made sure that all points of inflection (steeper places on cumulative curve) and sag (flatter places) are covered in the process. For very accurate work, a cumulative probability curve should be plotted first, and then from this probability curve, data points may be taken to construct a much more accurate cumulative curve on the arithmetic graph paper.

Statistical Parameters

Measures of Average Size

It is desirable to have a measure which will say, "Sample A is so much coarser than sample B". This is not as easy as it looks, though, because there are many different measures of average size. Using one measure, sample A might be "coarser"; using another measure, sample B might be coarser. The parameters which help us in taking decision are:

- Mode (M_O): Mode is the most frequently-occurring particle diameter. It is the diameter corresponding to the steepest point (point of inflection) on the cumulative curve (only if the curve has an arithmetic frequency scale). It corresponds to the highest point on the frequency curve.
- Median (M_d): Half of the particles by weight are coarser than the median, and half are finer. It is the diameter corresponding to the 50% mark on the cumulative curve and may be expressed either in Φ or mm. ($M_{d\Phi}$ or M_{dmm}).
- Graphic Mean (M_Z): The best graphic measure for determining overall size is the Graphic Mean, given by the formula $M_Z = (\Phi I6 + \Phi SO + \Phi 84)/3$.

Inclusive Graphic Standard Deviation (σ_G)

The Graphic Standard Deviation, σ_G, is a good measure of sorting and is computed as $(\Phi 84 - \Phi 16)/2$. However, this takes in only the central two thirds of the curve and a better measure is the inclusive Graphic Standard Deviation, σ_1, given by the formula:

$$(\Phi 84 - \Phi 16)/4 + (\Phi 95 - \Phi 5)/6.6$$

Measurement of sorting values for a large number of sediments has suggested the following verbal classification scale for sorting:

σ_1 under,

- .35 Φ, very well sorted,
- .35-.50 Φ, well sorted,
- .50-.71 Φ, moderately well sorted,
- .7l- l.0 Φ, moderately sorted,
- 1 .0- 2.0 Φ, poorly sorted,
- 2.0-4.0 Φ, very poorly sorted,
- over 4.041, extremely poorly sorted.

Measures of Skewness or Asymmetry

Curves may be similar in average size and in sorting but one may be symmetrical, the other asymmetrical. Skewness measures the degree of asymmetry as well as the "sign" - i.e., whether a curve has an asymmetrical tail on the left or right.

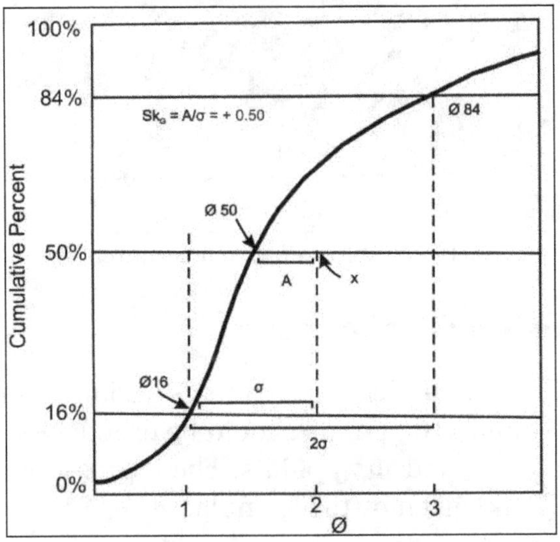

Measure of asymmetry (skewness) of curve.

Let "x" be the midpoint of the $\Phi 16$ and $\Phi 84$ values, found by: $(\Phi 16+\Phi 84)/2$.

In this case (1 +3) /2 or 2.0 Φ. Then the distance "A" is the displacement of the Median ($\Phi 50$) from the x midpoint.

The skewness measure is then A/σ,

Where, $A=(\Phi 16 + \Phi 84)/2 - \Phi 50$ and $\sigma=(\Phi 84 - \Phi 16)/2$,

So, Skewness (SK_G) is = $\Phi 16 + \Phi 84 - 2\Phi 50/(\Phi 84 - \Phi 16)$.

In this case, 1 + 3 - 2(1.5) / (3 - 1) Or SK_G = +0.50.

Inclusive Graphic Skewness (S_k)

The skewness measure discussed above covers only the central 68% of the curve. As most skewness occurs in the tails of the curve, this is not a sensitive enough measure. A much better statistic, one that includes 90% of the curve, is the Inclusive Graphic. This formula simply averages the skewness obtained using the $\Phi 16$ and $\Phi 84$ points with the skewness obtained by using the $\Phi 5$ and $\Phi 95$ points, both determined by exactly the same principle.

$$\left[(\Phi 16 + \Phi 84 - 2\Phi 50)/2(\Phi 84 - \Phi 16)\right]+\left[(\Phi 5 + \Phi 95 - 2\,\Phi 50)/2(\Phi 95 - \Phi 5)\right]$$

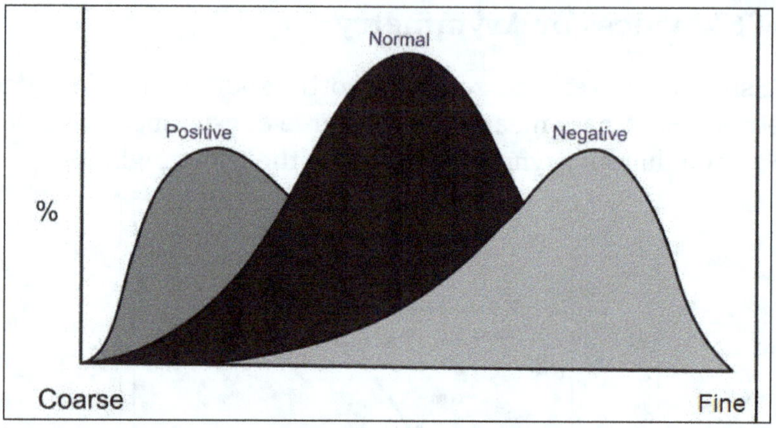

Grain size: Illustration of normal, positively skewed and negatively skewed distribution patterns.

Measures of Kurtosis or Peakedness

In the normal probability curve, defined by the Gaussian formula, the phi diameter interval between the Φ 5 and Φ 95 points should be exactly 2.44 times the phi diameter interval between the Φ 25 and Φ 75 points. The kurtosis measure used here is the Graphic Kurtosis, KG, (Folk) given by the formula:

$$K_G = (\Phi 95 - \Phi 5) / [2.44(\Phi 75 - \Phi 25)]$$

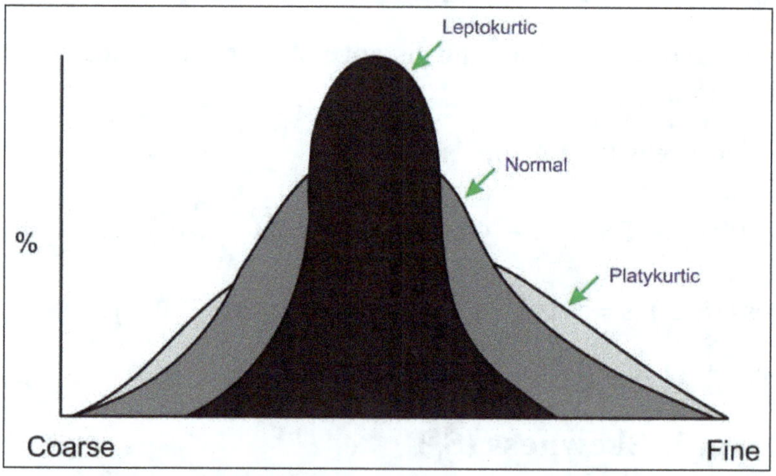

Illustration of peakedness in distribution patterns.

Significance of Grain Size Parameters

Grain size distribution curve is a result of interplay of many factors which include: 1) availability of grain sizes in the provenance, 2) sediment transport mechanism i.e. bed-load, saltation or suspension and 3) depositional environment and processes operative therein. Hence, a careful analysis of grain size distribution curve may offer understanding about each of these factors.

Control of Provenance

Depending on lithology, erosion and weathering history, specific grain sizes are supplied of any source area. Sediments generated out of chemical weathering of acid igneous rock (granites) are characteristically quartz grains corresponding the size of quartz crystals in granite and clay minerals (kaolinite, smectite, illite, etc.), principally developed through decomposition of feldspars in weathering profile. In contrast, basic rocks (gabbro, dolerite etc.) on weathering give rise to clay minerals and lithic fragments (where erosion is high) but practically no sand grains.

Transportation History

A grain size distribution curve may consist of several populations of grain sizes, each of which may have log-normal distribution and assume straight line when plotted on probability paper. Each segment of the curve is interpreted to represent different sub-populations of grains that were transported simultaneously but by different transport processes. Researchers have suggested distinction between sediments of different environments based on general shape of the curve, slopes of curve segments and positions of truncation points (breaks in slope) between straight line segments.

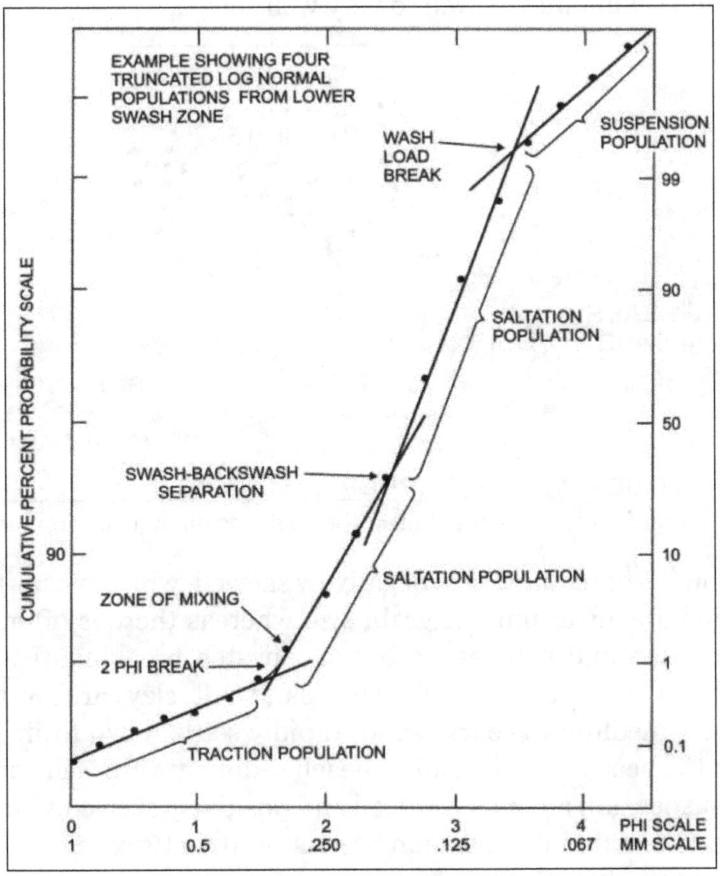

Distinction of size population.

However, a particular grain size distribution does not point unambiguously to a particular environment because of variability in depositional conditions within major environmental settings as well as due to operation of similar hydrodynamic conditions in different environments.

Depositional Environment

Fluvial sediments normally show poor sorting and positive skewness i.e. a wide spread towards the finer grain sizes (higher phi values) and a sharp delimitation at large grain-size end. For any river system there will be a reasonably definite upper limit of grain size that it can transport as bedload (competency), whereas there will be no sorting of finer size fractions that it can carry in suspension. In fact there is no process in rivers which preferentially remove finer grain sizes. Major variations in river flow velocity may occur during floods and cause poor sorting in river sediments.

Eolian (wind) sediments also have positive skewness because there is an upper limit of coarse grain size that wind can transport. However, there may be a tail of finer grain size in eolian dune sediment as fine grain sediments can be protected from erosion and transport under the armor of relatively larger sand-sized grains. However, unlike fluvial sediments eolian sediments are well to very well sorted.

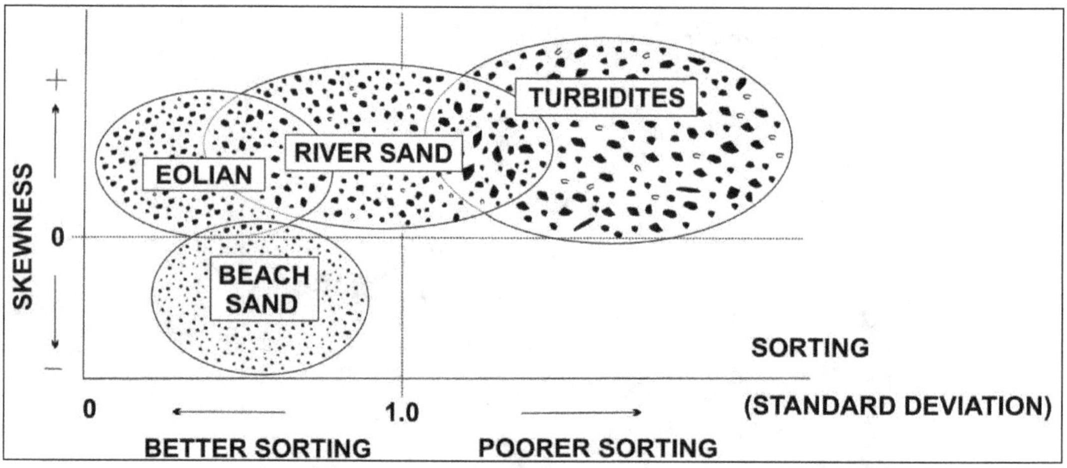

Granulometric characters of sediments belonging to different environments.

Beach deposits, on the other hand, are negatively skewed, which means the distribution curve shows a definite lower limit of grain size whereas there is often a tail of larger grain size i.e. granules and pebbles. In beach repeated breaking of waves take sediments in suspension and finer sediments (fine sand, silt, clay) are swept away. Coarse grains, particularly medium to coarse sand, rapidly settle down from suspension and get deposited on the beach again. Besides, beach sediments are well sorted. Sediments deposited from suspension are poorly sorted and positively skewed. Turbiditic currents in which sediments remain in suspension because of fluid turbulence allow suspension settlement and hence, characterized by poor sorting and positive skewness. Although this character is nearly similar with characters of fluvial sediments, distinction between these two products can be made on the basis of average grain size; average grain size of fluvial sediments much coarse compared to turbidite sediments.

Glacial transportation does not allow separation of different size grades and hence, tills (glacial deposits) have extremely poor sorting.

Particle Shape

Two parameters viz. roundness and sphericity are used to describe shape of any clastic grain.

Roundness

It is a property of surface shape i.e. whether a grain is smooth or angular. Roundness is defined as the sum of all (n) radii (r) of circles which can be inscribed by a section through the grain, divided by the radius (R) of the maximum size inscribed circle.

$$\text{Roundness} = \frac{\Sigma r / R}{n}$$

As it is difficult to measure in practice, a visual scale for degree of roundness of sedimentary grains as shown below:

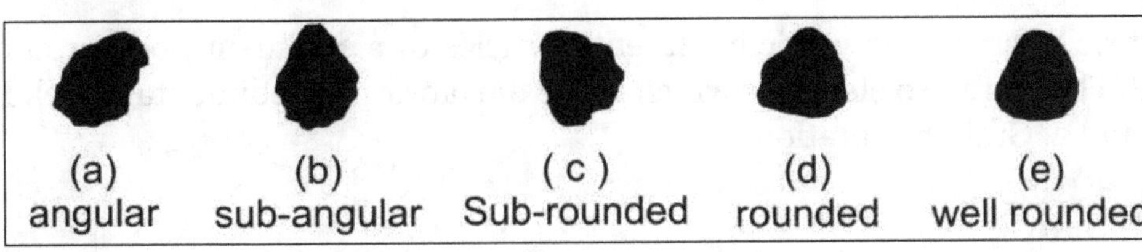

Sphericity

This is an expression for expressing degree of deviation of shape of a grain from an ideal spherical form. The parameter is defined as the ratio between the diameter of a circumscribed circle around the grain and the diameter of a sphere which has same volume with that of the grain i.e. nominal diameter. For an ellipsoidal grain with three axes of diameters d_L, d_I, d_S (longest, intermediate and shortest) the sphericity (ψ) is defined as:

$$\Psi = 3\sqrt{\left(d_s . d_I / d_L^2\right)}$$

For grains of any other shape, sphericity is defined as the ratio (s/S) between surface area of any sphere having same volume with that of a grain (s) and the actual surface area of the grain (S). Considering the difficulty in measuring surface area the ratio is replaced by the ratio d_n/D_s, where d_n is the diameter of a sphere having same volume with that of grain and D_s the diameter of circumscribing sphere i.e. the longest diameter of the grain. To simplify the process, sedimentologists have classified four shapes of grain based on b/a and c/b ratios, where a, b and c represents length, breadth and thickness of a grain, respectively. Four classes defined thereby are oblate, prolate, triaxial and equiaxial as described in the following diagram.

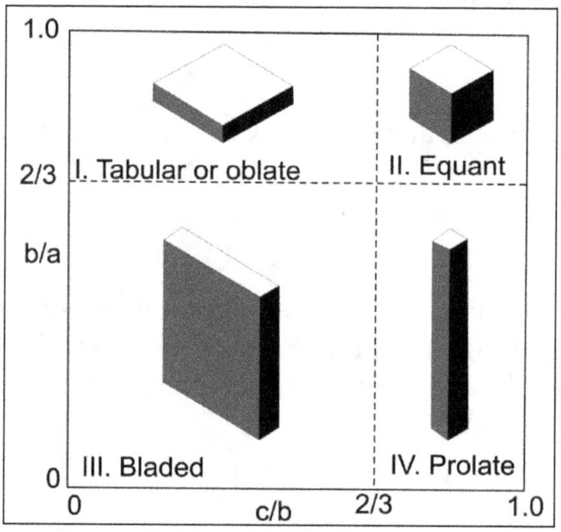

Grain Fabric (Packing and Orientation)

Grain fabric refers to the way constituent particles of a sediment body are mutually arranged. There are two elements which guide the fabric of a sedimentary rock 1. Grain packing and 2. Grain orientation.

Grain Packing

With spheres of uniform size six different packing motifs are described those vary from cubic (most loosely packed with a theoretical porosity value ~48%) to rhombohedral (most closely packed with theoretical porosity value ~26%). A more realistic packing value can be obtained taking prolate spheroids into consideration.

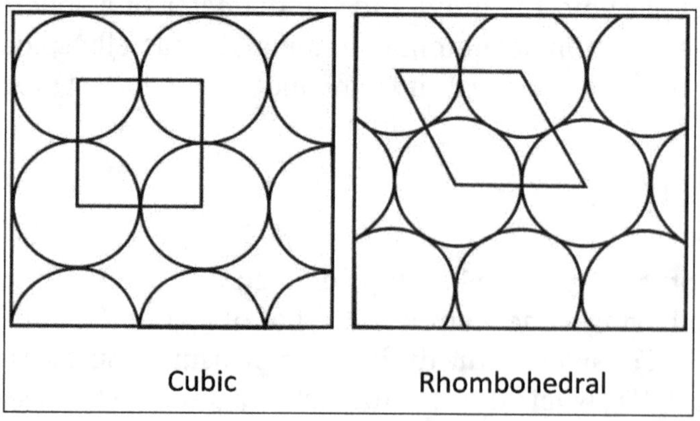

The 'packing index' is defined as the product of number of grain contacts observed in a thin-section traverse and the average grain diameter divided by the length of the traverse. However, post depositional compaction changes the grain shape and arrangement so much that depositional packing leave very little effect on porosity of lithified sediment.

Grain Orientation

Orientation of a grain is commonly referred with respect to the direction of flow in which it got deposited and the horizontal plane. Sand grain orientation is difficult to document because of difficulty in measurement. However, for pebble/gravel size clasts a very common orientation pattern is referred to as 'imbrication', which often is used as a good paleocurrent indicator. Two different imbrication patterns are noted i.e. 1) a(t) b(i) and 2) a(p)a(i); where 'a' refers to the longest axis of the pebble and 'b' refers to its intermediate axis. The a(t)b(i) imbrication (long axis transverse with the flow direction and intermediate axis dipping upcurrent) is noted in clasts transported as bed load, whereas clasts present within matrix- supported conglomerates, which are product of debris flows, can show presence of a(p)a(i) imbrication (long axis parallel with flow direction and dipping upcurrent).

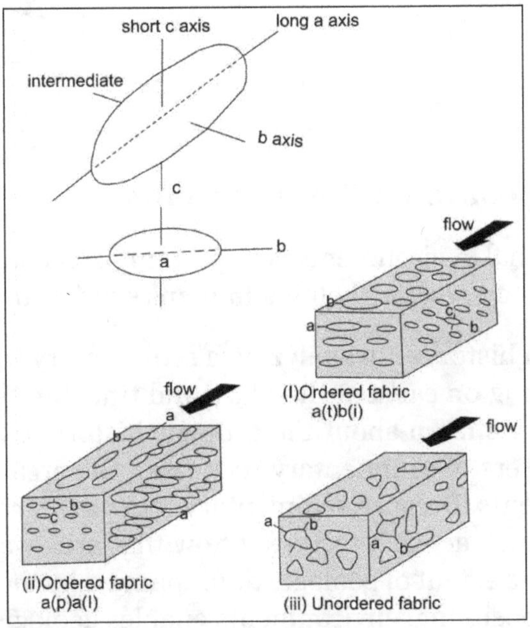

Depositional Environments

A depositional environment is a specific type of place in which sediments are deposited, such as a stream channel, a lake, or the bottom of the deep ocean. They are sometimes called sedimentary environments. The layers of sediment that accumulate in each type of depositional environment have distinctive characteristics that provide important information regarding the geologic history of an area. The characteristics that can be observed and measured in a sedimentary rock to deduce its depositional environment include its lithology (which is essentially its rock type), its sedimentary structures, and any fossils it may contain. The schematic diagram below shows different types of depositional environments.

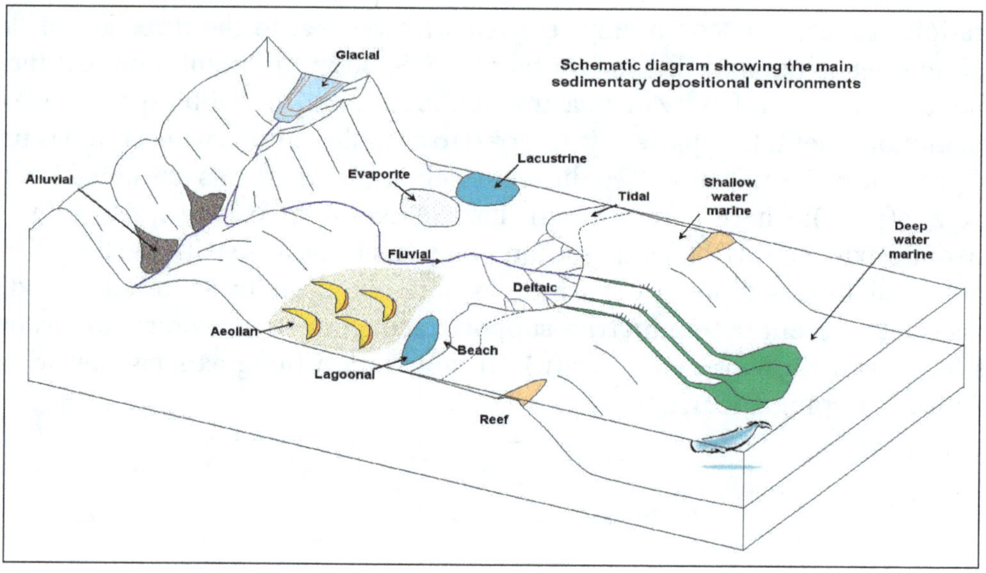

Importance of Depositional Environments

Knowledge of depositional environments is important for reconstructing earth history, understanding earth processes, and helping humans survive and prosper on earth.

1. Reconstructing earth history - By analyzing a sedimentary rock, a geologist can deduce what was happening on earth at the place and time the sediment was originally being deposited. What we learn about the geologic history of a region comes mostly from examining the layers of sedimentary rock from the area and determining their depositional environments. Because sedimentary rocks are stratified in age sequence, layers of sedimentary rock act as a record of how that area was changing, physically and biologically, over the extent of geologic time spanned by the sedimentary rock layers. Reconstructing depositional environments enables geologists to observe climates of the past, life forms of the past, and geography of the past -- the location of mountains, basins, large rivers, and bays of the ocean. Changes over time in climate, life forms and geography constitute the geologic history of a region. Ultimately, regional geologic histories are compiled into a history of the earth over the whole course of its existence, including the formation, growth, and movements of continents and ocean basins, the growth and erosion of major mountain ranges, and the history of life on earth.

2. Understanding earth processes - Sediments are deposited in many environments on the earth's surface, some of which humans have little familiarity with, such as deep ocean environments. Sediments have been deposited in the past in environments that do not exist in the present, such as an atmosphere with no free oxygen, or an environment disturbed catastrophically by a gigantic meteorite impact. Therefore, by examining sedimentary rocks as windows into these environments, we can learn about earth processes that we would otherwise know little about, and deduce details about them such as the chemistry of the air or water with which the sediments were in contact and the physical processes that were occurring in that environment.

3. Helping humans survive and prosper - Knowledge of earth processes has many practical applications for human health and survival. To give one example, by reconstructing depositional environments of certain sediments deposited along the coast of the Pacific Northwest, geologists concluded that great subduction earthquakes and tsunamis (giant waves) created by the earthquakes, were the driving forces of the depositional environment of those sedimentary deposits. This has led to re-evaluation of the earthquake hazards in western Washington and Oregon and rewriting of building codes and engineering standards for construction of schools, roads, bridges, and infrastructure in that area. This has affected such things as insurance policies and construction costs.

Geologists use analyses of depositional environments to help locate, inside the earth, sources of oil, coal, natural gas, deposits of valuable metals/minerals/rocks, and aquifers, which are useable sources of groundwater.

The many depositional environments which can be grouped into three major categories - marine, transitional, and continental.

Identification of Depositional Environments

The characteristics of a sedimentary rock that are affected by its depositional environment are its sedimentary lithology (the minerals and texture of the rock), its sedimentary structures, and its fossils. Sedimentary rocks contain sedimentary structures that were formed as the sediments were being deposited. Many sedimentary rocks also contain fossils, which are our main source of information of the history of life on earth. Sedimentary structures, and fossils, are best found and examined in outcrops, where whole beds of sedimentary rocks are exposed in their undisturbed geological setting. The structures and fossils in sedimentary rocks reveal what was happening on the earth at the place and time the sediments were being deposited.

Lithology

Sedimentary lithology is a combination of the mineral content and sedimentary texture of the rock. The lithology of a sedimentary rock is largely summarized in the name of the rock.

Sedimentary Structures

Sedimentary structures such as cross-beds, graded beds, and mud cracks are useful for determining which way was up in the original sequence of sediments. It is possible for tectonic forces to deform rocks in the crust to the point that beds of sedimentary rocks have been turned upside down. Therefore, a geologist needs to check the sedimentary structure to be sure which way was up, especially if looking at beds of sediment that have been tilted to high angles, far from their original horizontal position.

Bedding

Sedimentary beds, or strata, are layers of sediment that can be distinguished from layers above or below by the type, texture, or color of the sediment. Most sediment accumulates under water on the surface of the earth. Some accumulate on the earth's surface at the base of the atmosphere. In either case the deposition of sediment tends to occur in events or pulses of increased sedimentation, such as during high flows or floods of rivers, seasons of strong wind in a desert, certain parts of the tide cycle in shallow marine environments, or yearly freeze and thaw cycles in lakes in sub-arctic environments. The result is sedimentary beds that may be only a few mm thick or may be up to several m thick. Note that the processes that caused the bedding may be inferred, with careful study, from the nature of the bedding itself.

Bedding thickness - The thickness of sedimentary beds can be measured and described using standardized terminology as follows:

- Massive: No beds apparent over several m of exposure.
- Thick beds: greater than 100.0 cm (greater than 1.0 m).
- Moderate beds: 10.0 cm - 100.0 mm.
- Thin beds: 1.0 cm - 10.0 cm.
- Laminae or laminated beds: less than 1.0 cm (less than 10 mm).

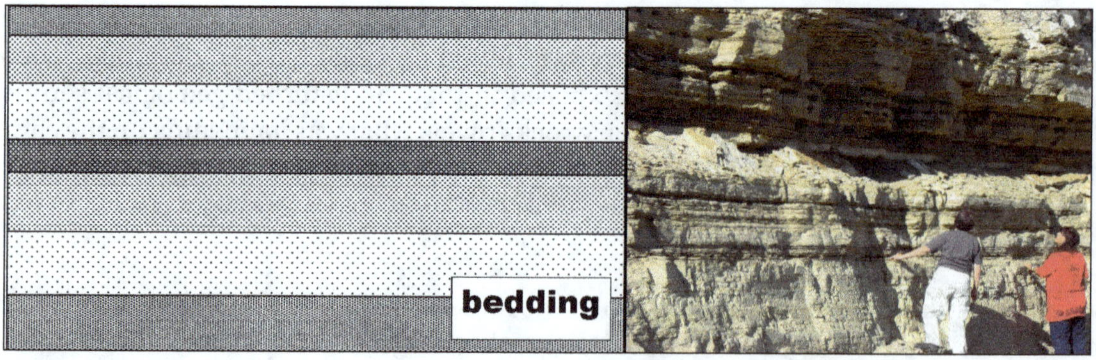

Cross Bedding

The fact that in many cases sediments tend to settle from water (or air) and fill in low areas with relatively flat layers is the basis of the principle of original horizontality, one of the important principles of relative geologic age. However, not all sedimentary beds are horizontal to begin with. Cross-beds in particular begin as inclined beds, formed by sediment piling up in layers on the slopes of sediment ripples or dunes, or on slopes that go gradually into deeper water as sediments pile up from a river's mouth into an ocean or lake. Cross-beds formed from sediment ripples being moved at the base of a current of water slope downward in the direction the water was flowing. Wind-blown sediment that were deposited in the form of sand dunes form long cross-beds that represent the migrating, down-wind faces of the sand dunes.

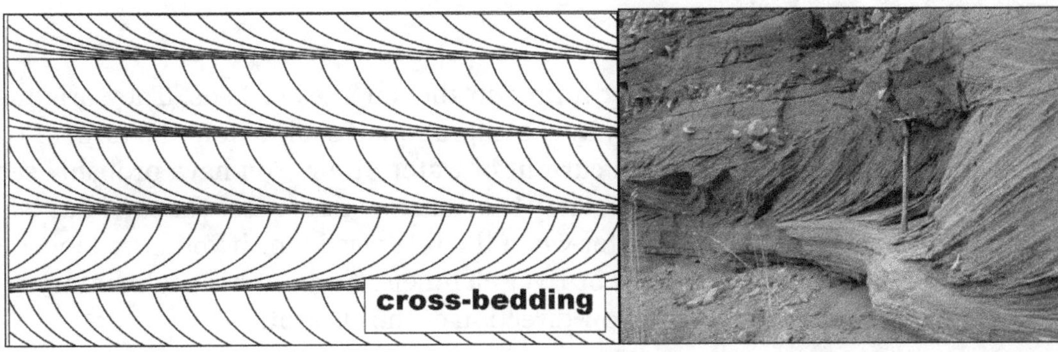

Rhythmic Bedding

Rhythmic bedding consists of a repeated sequence of beds. Varves are a simple example of rhythmic bedding. Turbidites are a more complex example of rhythmic bedding. Rhythmic beds are sometimes called "rhythmites."

Graded Bedding

Graded beds have coarser (larger) sediment grains at the bottom, grading up to finer (smaller) sediment grains at the top of the bed; or the grading may occur in a sequence of beds from, at minimum, a bed of coarse sediment overlain by a bed of finer sediment, or several beds of finer and finer sediment on top of each other. Graded bedding results from the fact that larger grains of common rocks or minerals fall out of a body of water faster than the finer grains of sediment do. Once a flow of water slows enough for the sediment grains to settle out, if the sediment grains are in a mixture of sizes, they will form a sedimentary bed, or continuous sequence of sedimentary beds, with the larger sediment grains at the bottom and the larger sediment grains at the top.

Ripples

Sediment ripples are a structure that forms on the surfaces of beds. They originate similar to the way cross-beds develop, by the migration of sediment in the form of ripples, or larger dunes, at the base of a current of water or air. You have probably seen sediment ripples if you have been to a sandy beach at low tide where the sand has been formed into ripples by the flow of water when the tide was in, or if you have looked at sandy sediment at the base of stream or river channels. Asymmetric sedimentary ripples have steeper faces in the down-flow direction of the current.

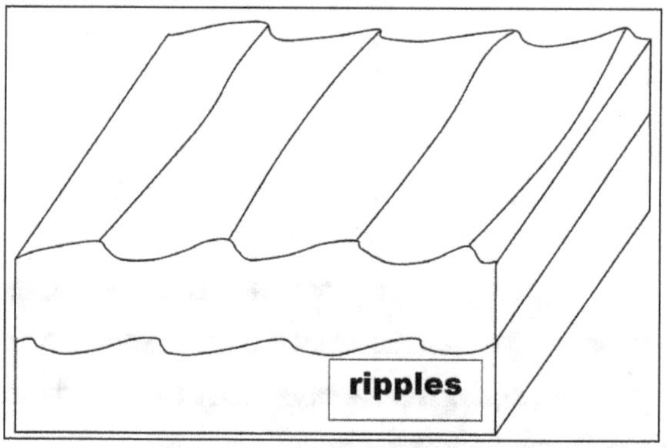

Mud Cracks

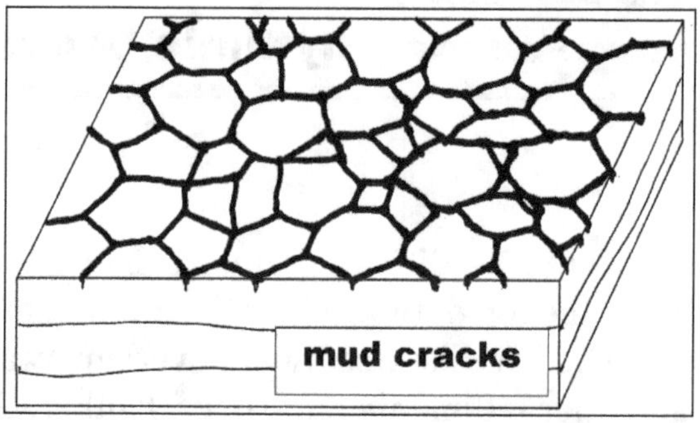

Fine-grained sediment, particularly sediment composed at least partly of clay, will form a polygonal pattern of mud cracks on the surface of the bed, if the sediment was covered by water which dried up or receded and left the bed exposed to the air.

Fossils

Fossils are the remains or traces of biological organisms preserved in rocks. Fossils are commonly found in sedimentary rocks. Besides providing evidence of life forms that have existed in earth's past, and how life on earth has evolved over the course of earth history, fossils provide important information about the depositional environment in which the sediments were deposited. For example, fish fossils imply that the sediments were deposited in a body of water. Fossils of the leaves of trees imply that the sediments were deposited on land, above sea level.

Trace fossils, such as dinosaur footprints preserved on a lithified stratum of mud, or wormholes in silt from the floor of a shallow sea are another important type of fossils. Wormholes or tunnels created by other organisms that lived in the sediment and are preserved as trace fossils are known as bioturbation.

The original organism is likely to be preserved only in terms of its hard parts, such as its shell, skeleton, or teeth. Soft parts of animals are much less likely to be fossilized. Because a certain combination of events and conditions is required for dead organisms to become fossilized, most organisms that live on earth never become fossilized. Many whole species have existed that are not represented in the fossil record. However, there are more fossils in the rocks that have yet to be discovered. Paleontologists - professionals who study fossils scientifically - may occasionally have the pleasure of confirming and reporting the discovery of a previously unknown fossil species.

Fossils are most commonly found in sedimentary rocks. Less commonly, fossils occur in certain types of volcanic rocks, rarely in low-grade metamorphic rocks, and never in plutonic igneous rocks or high-grade metamorphic rocks. Not all sedimentary rocks will contain fossils, but many do, which adds greatly to the information the rock contains about its depositional environment and what it represents in terms of the history of life on earth.

Fossils may not preserve any of the original tissue, bone, or shell that the organism was composed of. Once buried in the earth as part of the rock cycle to become lithified, a fossil may become mineralized completely. For example, petrified wood has had its organic material replaced by quartz as a result of chemical reactions that occurred once it was buried deep enough to be below the water table. Sometimes shells or exoskeletons of marine animals are found to have been completely replaced by the shiny yellow mineral pyrite, forming a detailed mold of the original shell or exoskeleton.

Sedimentary Facies

Sedimentary facies are bodies of sediment that originate simultaneously in adjacent depositional environments. For example, a beach facies can usually be distinguished from a tide flat facies, both of which were deposited at the same time adjacent to each other. Compared to the beach facies, the tide flat facies will have smaller average sediment grain size, more bioturbation fossils, contain cross-beds and ripples created by tidal currents, and have more mollusk or other shallow-water fossils preserved in their original place, in unbroken form. There will not be a sharp boundary between the two facies preserved in the sedimentary record. Instead the boundary between them will be a zone with beds of sediment that interfinger and grade into each other sideways from one facies to another.

Below is a simplified diagram of three sedimentary facies adjacent to each other: a beach and tide flat facies (combined) a marine or near-shore portion of a continental shelf, and an offshore carbonate platform or reef. The beach and tide flat facies sediments are mostly sand, the bay facies is mostly mud, and the reef facies is mostly shells and corals which are made of carbonate minerals. If these sediments are buried and lithified into sedimentary rocks, the beach sands turn into sandstone, the bay mud turns into shale, and the reef sediments turn into limestone.

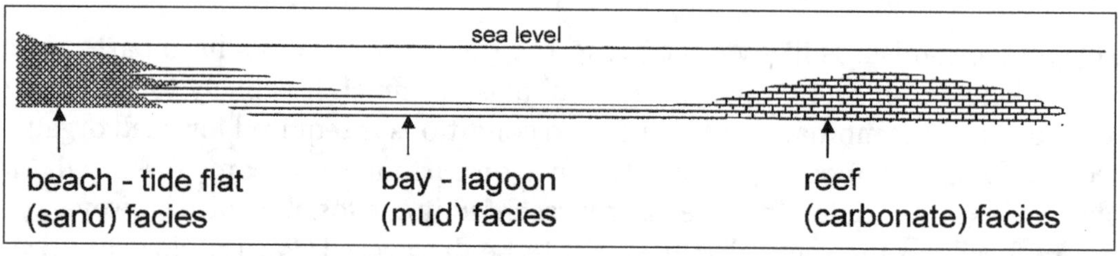

The study of sedimentary facies has revealed, among other things, how sea level, relative to the shore of a continent, is constantly changing over the course of geologic time, on time scales that can vary from decades to millions of years. To give a more specific example of how facies changes record sea level change, deep in the Grand Canyon of Arizona is a sequence of three sedimentary rock formations: the Tapeats Sandstone, the Bright Angel Shale, and the Muav Limestone. Those three sedimentary formations are thought to have originated as a continuous series of sediment as sea level gradually rose, relative to the land, over a span of time that took over a million years in that area.

As sea level grew deeper, the shore of the ocean moved inland, which means that the beach facies moved inland, the bay facies shifted in the same direction, and so did the reef facies. Along the low-gradient coast of the continent, as sea level rose higher, what had been a beach was covered by deeper water and became the bottom of a bay where mud accumulated on top of the sand. Then, as sea level rose higher still, the area was in deeper water farther from shore where the water was relatively clear and free of clastic sediments. This allowed a coral reef to build on top of the mud. A more geological time passed and the environment of the area changed again, the deposits of sand, mud, and carbonate sediment were buried and lithified into the sequence of sedimentary formations which, from bottom to top, are sandstone, shale, and limestone. The sequence of sediments that record a gradual sideways shift of sedimentary facies during a marine transgression is shown in the diagram below. In the diagram, the part of the continent above sea level would be on the left. As time passed the shoreline was moving to the left.

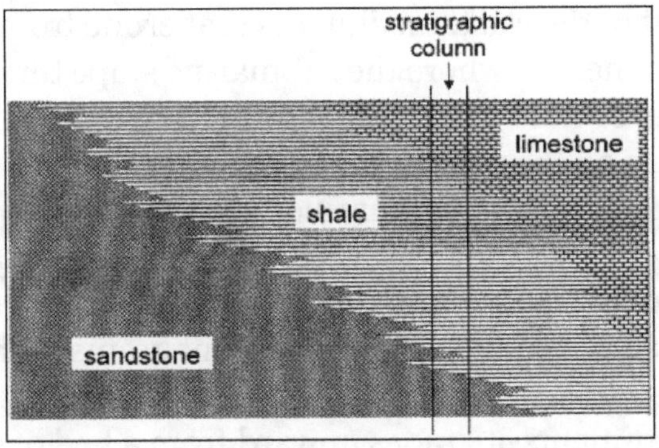

At a given location, such as where the Grand Canyon is now located, evidence of a marine transgression appears as a continuous stratigraphic sequence of sandstone at the bottom, shale above the sandstone, and limestone on top of the shale. The minerals, sedimentary textures, sedimentary structures, and fossils are specifically indicative of beach, tide flat, muddy bay, and offshore reef depositional environments. Such a transgressive sequence is marked as a stratigraphic column. Below is shown a simplified version of the stratigraphic column that represents a marine transgression, with the oldest sedimentary formation at the bottom.

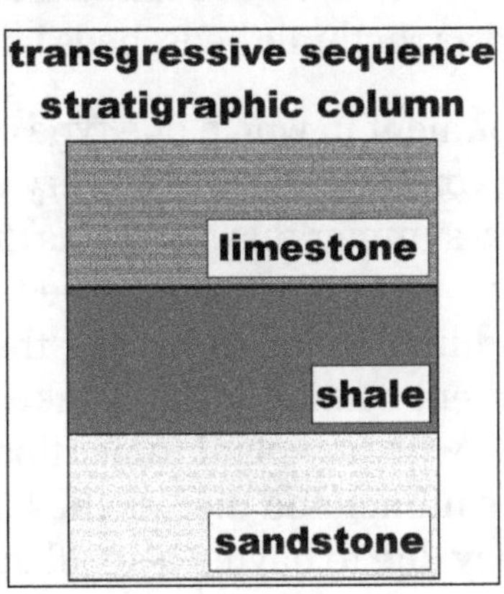

It is also possible for a regressive sequence to occur as sea level goes down relative to the coast of a continent, resulting in the opposite sequence: limestone on the bottom, shale in the middle, and sandstone on top. Regressive sequences are less likely to be preserved in the rock record than are transgressive sequences. This is because, as sea level falls, the exposed parts of the continent, which had previously been below sea level, are exposed above sea level and more subjected to the forces of weathering and erosion. Therefore the sediments are likely to be removed by earth processes rather than kept buried and preserved within the earth.

Examples of Sediments and their Particular Depositional Environments

Turbidites

The oceans receive most of the clastic sediments that erode from the continents. On the edges of the continental shelves, where the submarine slope tilts down into much deeper water, accumulations of mud and sand deposited by rivers build up. Eventually so much sediment builds up on the edge of the steepening slope that it is likely to give way into an underwater landslide. The submarine landslide will flow down the slope into deeper water, mixing with seawater as it goes to form what is called turbidity current. As the sediments gradually settle out of the turbidity current onto the deeper ocean floor, the coarser-grained sediments (those sediment grains with larger diameters) will settle to the bottom first, followed gradually by finer and finer sediments. This creates a graded sequence of sediments-it grades upward from a bed of sand through a layer of silt to a top layer of fine mud. This graded deposit becomes a rock known as a turbidite. Over the years one turbidite is likely to be deposited on top of another, over and over again thousands of times. This creates repeated beds of coarse sand to fine mud, which may total thousands of feet thick. If parts of the ocean floor end up becoming part of a continent, turbidites are likely to be a major component the accreted terrane.

Varves

Varves are annual layers of sediment, layers of sediment that accumulate each year, year after year. Varves are deposited as rhythmic beds, beds laid down in a repeating pattern.

A common depositional environment in which one type of varve is deposited is lakes in cold climates where the surface of the lake freezes every winter and thaws every spring and summer. During the spring-summer thaw, streams discharge at a high rate into the lake, causing the deposition of a layer of silt on the bed of the lake. The silt is usually rich in quartz and feldspar and light-colored. During the winter freeze, when there is little or no stream-borne sediment coming into the lake, only clay-size particles settle to the bottom of the lake, along with any planktonic (floating, mostly microscopic) organisms that flourished in the summer and died as the lake froze. The winter sediment is thus clay, sometimes dark clay due to having a small amount of carbon in it.

The resulting varve is a pair of strata: a light-colored stratum of silt from the spring-summer warm season, and a darker stratum of clay from the winter freeze.

Sequences of varves are especially common in locations that were the beds of lakes near glaciers during ice ages. Ice ages are times when continental glaciers formed and advanced outside of Polar Regions. The most recent ice age, the Pleistocene epoch (approximately 2.5 million to 12,000 years ago), saw continental glaciers advance several times in northern North America (into what is now the northern-most United States), the Scandinavian Peninsula and nearby parts of Europe including Britain, and parts of northern Asia. During continental glaciations, the glaciers dammed many stream drainages and created temporary lakes in cold climates next to the glaciers, where sequences of varves accumulated.

Marine Limestone

Limestone, rock made of the calcium carbonate mineral known as calcite, can form in a variety of depositional environments, from hot spring deposits in lakes to coral reefs in the tropical oceans. Most limestone originates in shallow waters of tropical oceans, and may carry fossils of plants and animals that lived in those marine environments. However, limestones made of buried coral reefs are not as common as limestones made simply from lime mud. Lime mud originates from disintegrated organisms that have hard parts made of calcium carbonate. As a result, limestone is commonly massive (lacks obvious beds), fine-grained, and lacks obvious fossils.

Tsunami Deposits

When devastating subduction zone earthquakes occur along a coast, extremely large water waves called tsunamis are generated. At the same time, sea level changes relative to land level along the local shore. The combination of a sudden drop in land level and a tsunami washing over coastal lowlands creates several distinctive markers in the sediment layers that remain. These include muddy coastal marsh deposits overlain by gravel or sand deposits that have sedimentary structures indicating high-energy waves flowed inland along the coast. Where the coast is nearly flat rather than steep, these tsunami deposits can extend miles inland. Groves of cedar trees or other evergreen trees that grow adjacent to marshy areas, barely above sea level, may drop down and have their tree roots subjected to salty water. This will kill the trees, though they may stand in place for several hundred years as "ghost forests," silent testimonials to great earthquakes of the recent past.

Coal

Coal is a chemical sedimentary rock made mostly of carbon. It forms from the remains of plants that lived in moist environments rich in trees, shrubs, water, and mud. In such swampy settings, the dead plant debris is quickly buried and thus escapes rotting away at the earth's surface. Upon being buried, heated and compressed within the earth's crust, the dead plants will become coal if the right conditions of heat and pressure are achieved.

Meandering Rivers

Sequences of beds of sandstone, conglomerate, siltstone, shale, and plant fossils indicate sediment deposition by a system of meandering rivers. If there were thick woods and densly vegetated swampy areas, there may also be coal. Details in the sedimentary structures, characteristic signatures of particular depositional processes, will confirm if there were meandering river channels, sandbars, stream bank erosion, and occasional floods.

Deltas

Much sediment is deposited where rivers empty into lakes, or into the ocean. This is because the velocity of the stream current comes to a stop there, and as the flow slows down, the sediments being transported by the stream settle to the bottom and are deposited. Deltas along ocean coasts are transitional environments, where the surface of the earth gradually slopes from on land to beneath the ocean, and where currents of fresh water and sediments eroded from continents meet waves, tides, and marine sediments.

Deltas where large rivers meet the ocean are huge, especially when their submarine parts are taken into account. The southern part of the state of Louisiana is on the Mississippi River delta. Beneath the Gulf of Mexico, there is a much larger volume of the delta sloping down to its base in deep water far from shore. The delta of the Brahmaputra River in Bangladesh is the subaerial part of a large delta that has a submarine component, known as the Bengal fan, which may be the largest body of sediment on earth.

It is common for oil deposits to be found in the sedimentary beds of deltas, including deltas from rivers that have long since disappeared, the sedimentary beds preserved as layers of sedimentary rock. Drilling into delta deposits by oil companies has led to detailed knowledge of the structures, minerals, textures, facies, and fossils that are typically deposited in different parts of a marine delta. Below is a simplified diagram of the major sets of beds that characterize a delta.

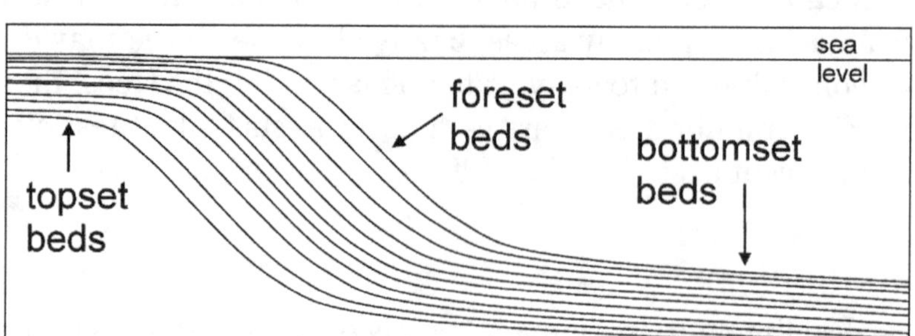

The foreset beds of a marine delta are sediments deposited in a continental setting, on the low-gradient parts of the delta above sea level, where there were meandering stream channels and marshy or swampy floodplains. The sediments tend to be fine-grained, thin-bedded, and have certain types of cross-beds, ripples, plant fossils, and in some cases mud cracks.

The foreset beds were deposited on higher-gradient slopes going down into deep water, so foreset beds consist of sediments deposited underwater in relatively high-energy conditions. The coarser sediments of turbidity currents - sand and gravel with cross-beds and graded beds - are common in the foreset beds.

The bottomset beds formed where turbidity currents, which originated higher on the delta, spread out onto the lower-gradient ocean floor in deeper water and lost their energy, and consist of fine sands, silts and clays, commonly in characteristic sequences of graded beds known as turbidites.

Sedimentary Basin

The official definition of a sedimentary basin is: a low area in the Earth's crust, of tectonic origin, in which sediments accumulate. Sedimentary basins range in size from as small as hundreds of meters to large parts of ocean basins. The essential element of the concept is tectonic creation of relief, to provide both a source of sediment and a relatively low place for the deposition of that sediment.

A sedimentary basin doesn't have to be a place on the Earth's surface with strictly basinal shape, with closed contours, like a washbowl: great masses of sediment can be deposited on a surface with a gentle and uniform slope. But implicit in the concept of a sedimentary basin is the existence of prolonged crustal subsidence, to make a place for a thick deposit of sediment that might well have been deposited in an area without basinal geometry at the surface. Tectonics is needed to make sedimentary basins, but the record of the basin itself is sedimentary.

As with most blanket statements, this one has exceptions to it. A sedimentary basin can be made just by erecting high land in an adjacent area by volcanism.

The term "sedimentary basin" is usually not applied to relatively thin and very extensive deposits of sandstone, limestone, and shale from epicontinental seas on the cratons, many of which have seen no deformation through billions of years, but only to relatively thick deposits in tectonically active areas with negative relief. (But intracratonic basins are the exception in this regard.)

Tectonics and Sedimentation

Tectonics is the most important control on sedimentation; climate is a rather distant second. The important effects of tectonics on sedimentation, direct or indirect, include the following:

- Nature of sediment.
- Rate of sediment supply.
- Rate of deposition.
- Depositional environment.
- Nature of source rocks.
- Nature of vertical succession.

In fact, tectonics affects climate itself, by way of effects as broad as the distribution of oceans and continents, and as local as rain shielding by local mountain ranges. And sedimentation itself affects tectonics, although to a much lesser extent, mainly by increasing the lithospheric loading in the basin.

The other side of the coin is that by far the best way of telling paleotectonics is by the sedimentary record in sedimentary basins. The disposition of sediment types, sediment thicknesses, and paleocurrents in a basin gives evidence of the existence and location of elevated areas of the crust created by tectonism.

The only basins that are preserved in their entirety are those that lie entirely in the subsurface! Basins exposed at the surface are undergoing destruction and loss of record by erosion. So there's an ironic tradeoff between having more complete preservation in the subsurface but less satisfactory observations.

How do you gather data on sedimentary basins? There aren't many ways, really: surface mapping; cores; and subsurface geophysics, mainly seismic profiling.

What kinds of things can you do with the data, Here's a list of some the fairly standard things you can do. These ranges from very descriptive to very interpretive. It makes sense to do the descriptive things first and then work toward the more interpretive.

- Master cross sections: With the present land surface as the most natural datum, construct several detailed physical cross sections through the basin to show its geometry and sediment fill.
- Stratigraphic sections: Construct a graph, with time along the vertical axis, showing the time correlations of all the major rock units along some generalized traverse across the basin. Such a section includes hiatuses, during which there was nondeposition or erosion.
- Isopach maps: With some distinctive stratigraphic horizon near the top of the

section as datum, draw a contour map showing isopachs (isopachs are loci of equal total sediment thickness) in the basin.

- Lithofacies maps: For one or a series of times, draw a map showing distribution of sediment types being deposited at that time.
- Ratio maps: Compute things like sand/shale ratio, integrated over the entire section or restricted to sometime interval, and plot a contour map of the values.
- Paleocurrent maps: For one or a series of times, draw a map showing the direction of paleocurrents in the basin at that time.
- Grain-size maps: For the entire basin fill, averaged vertically, or for some stratigraphic interval or time interval, draw a map that shows the areal distribution of sediment grain size. This is especially useful for conglomeratic basins.

Another interpretive thing you can do is draw diagrams (qualitative or semiquantitative) showing the evolution of depositional–paleogeographic– paleotectonic setting of the basin, by means of maps and sections. These could range from cartoons to detailed representations drawn to scale.

Just within the past ten years, computational techniques known as backstripping have been developed to "undo" the deposition in a sedimentary basin. This involves restoring the basin to a whole series of past configurations by removing one layer of sediment at a time and adjusting for compaction, subsidence, and sea-level change. This lets you reconstruct the configuration of the basin through time, perhaps by drawing palinspastic cross sections for various time intervals. In a way, this is the next best thing to having in your possession a time-lapse movie of the entire development of the basin.

This is a good place to warn you about vertical exaggeration of cross sections of sedimentary basins. Cross sections are almost always drawn with great vertical exaggeration, typically somewhere between 10:1 and 100:1. This is because in true scale most basins are relatively thin accumulations, hundreds to thousands of meters of sediment spread over distances of tens to hundreds of kilometers. So to see the relationships adequately in cross sections, the sections have to have great vertical exaggeration. Carefully constructed sections show both the vertical and the horizontal scale, but cartoons often don't show the scales.

Paleocurrents

Much effort has gone into developing ways of figuring out paths of dispersal of sedimentary material in basins. One of the standard ways is to measure paleocurrent directions recorded locally in the rocks. (A paleocurrent is just what the term implies: a current, of water or wind that existed at some time in the past.) Techniques are well established.

Knowledge of paleocurrents is helpful in solving both local and regional problems of sedimentary basins. Locally, paleocurrent directions can help you to figure out or predict, indirectly, the shape and orientation of sediment bodies, like channel sandstones. This has obvious advantages in petroleum exploration. Regionally, paleocurrent directions can help establish paleoslope and source of sediment supply to the basin.

Here's a list of the most important, with annotations:

- Cross-stratification: Measure the local orientation of laminae in the cross sets, on the theory that the local downdip direction, which presumably is the direction of progradation of the foreset slope, is likely to represent fairly closely the local current direction. That's true, however, only if the bed forms were reasonably two-dimensional. If the bed forms were threedimensional, resulting in trough cross stratification, measurement of foreset dip directions at local places in the cross sets can be very misleading; it's much better to try to ascertain the orientation of the trough fills themselves, although it takes good outcrops to do that. Seeing rib and furrow is by far the most reliable way of obtaining a paleocurrent direction from cross-stratified deposits, but unfortunately it's uncommon to see on outcrop.

- Bed forms: If you are lucky enough to see bedding planes covered with symmetrical ripples or dunes, you can get an excellent measurement of current direction.

- Clast orientation: Long axes of the larger clasts in a clastic deposit, whether gravel or sand, are commonly oriented by the current, although the orientation may be rather subtle. The problem is that the orientation relative to the current (flow-transverse? flow-parallel?) depends on the flow itself in ways not well understood. So beware of clast orientation in and of itself. Pebble imbrication is an exception, and should always be sought in gravels and conglomerates.

- Sole marks: Flutes and grooves at the bases of turbidites and other strong current-event beds give excellent evidence of the direction of the initial, eroding current. But keep in mind that the later current that did the depositing did not necessarily flow in exactly the same direction.

- Parting lineation: Parting lineation is thought to reflect a subtle anisotropy in rock strength caused by a statistical tendency toward alignment of sand grains in sandstone parallel to the current direction. It gives excellent evidence of the orientation of the current, but unfortunately not the direction.

The paleocurrent measurements you take from dipping beds are no good in themselves: what you need to do is "undeform" the strata by rotating them back to horizontal, taking your paleocurrent measurements with them. That's straightforward provided that the strata are not strongly deformed. But the greater the deformation, the more uncertain is the exact way you should be undeforming the strata.

How Basins are Made

In one sense, the origin of sedimentary basins boils down to the question of how relief on the Earth is created. Basically, there are only a few ways.

Local

On a small scale, hundreds to thousands of meters laterally, fault movements can create relief of hundreds to thousands of meters, resulting in small but often deep basins (some of these are called intermontane basins; think about places like Death Valley). You might guess that it takes dip-slip fault movements to create new relief, but that's not true: steps (in the proper sense) along strike-slip faults can produce small pull-apart basins. Relief of this kind is on such a small scale that it tends not to be isostatically compensated. It's like setting a block of granite out on your driveway; the flexural rigidity of your driveway is great enough compared with the imposed load that the granite block is prevented from finding its buoyant equilibrium position.

Regional

Basin relief can be created mechanically on a regional scale in two very important ways: thermally or flexurally, or by a combination of those two effects). Keep in mind that basins can also be made just by making mountain ranges, on land or in the ocean, by volcanism.

Thermal

If the lithosphere is heated from below, it expands slightly and thus becomes less dense. This less dense lithosphere adjusts isostatically to float higher in the asthenosphere, producing what we see at the Earth's surface as crustal uplift. If the lithosphere cools back to its original temperature, there's isostatic subsidence back to the original level.

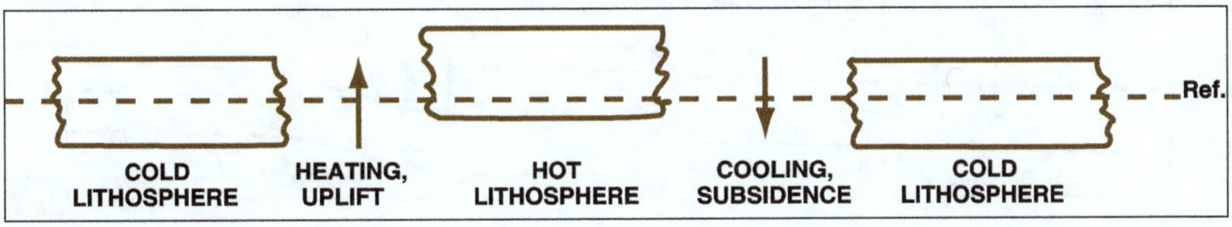

But suppose that some erosion took place while the crust was elevated. The crust is thinned where the erosion took place (and thickened somewhere else, where there was deposition; that might be far away, at the mouth of some long river system), so when the crust cools again it subsides to a position lower than where it started, thus creating a basin available for filling by sediments.

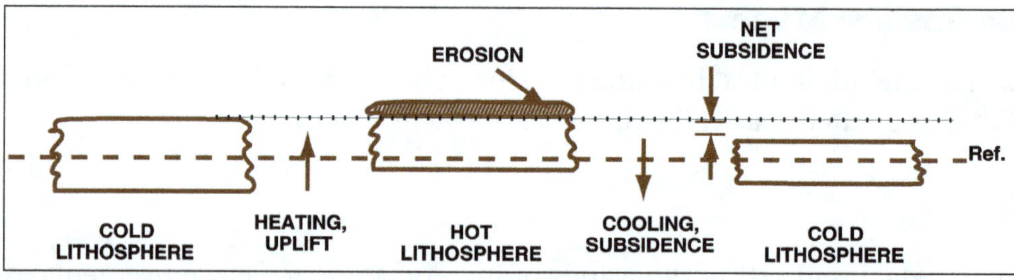

But the magnitude of crustal lowering by this mechanism is less than is often observed in basins thought to be created thermally. It has therefore been proposed, and widely accepted, that in many cases extensional thinning of the lithosphere accompanies the heating. Then, upon recooling, the elevation of the top of the lithosphere is less than before the heating and extension. This kind of subsidence has been invoked to explain many sedimentary basins.

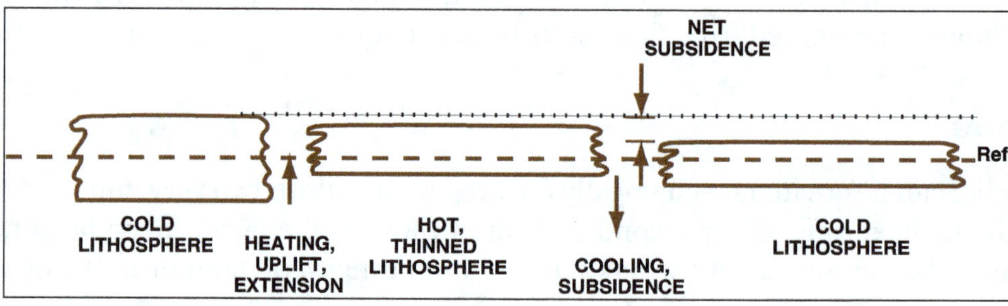

Flexural

Another important way to make basins is to park a large load on some area of the lithosphere. The new load causes that lithosphere to subside by isostatic adjustment. But because the lithosphere has considerable flexural rigidity, adjacent lithosphere is bowed down also. The region between the high-standing load and the lithosphere in the far field is thus depressed to form a basin. This model has been very successful in accounting for the features of foreland basins, which are formed ahead of large thrust sheets that move out from orogenic areas onto previously undeformed cratonal lithosphere.

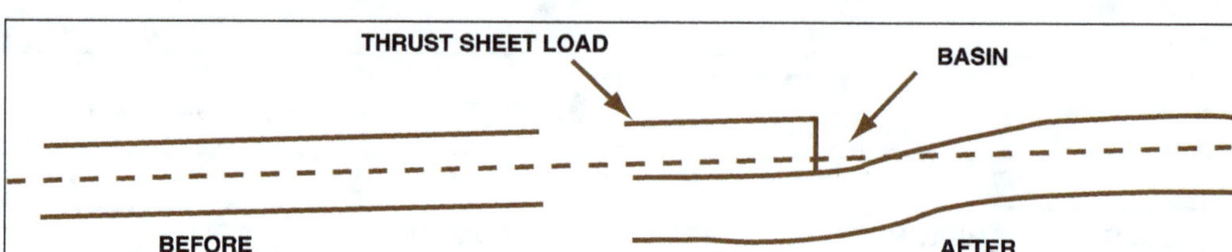

Geosynclines

The concept of geosynclines was developed in the last century to deal with the existence of thick successions of sedimentary rocks in what we would today call orogenic belts. A geosyncline is large troughlike or basinlike downwarping of the crust in which thick sedimentary and volcanic rocks accumulated. Usually, but not always, such accumulations are deformed during a later phase of the same geological cycle in which they were deposited. There is a close, although not one-to-one, correspondence between geosynclines and what we are discussing here as sedimentary basins.

The geosyncline concept was developed in an effort to understand the regularities of sedimentation in orogenic belts. Over the decades, in both Europe and North America, the concept was elaborated to an extreme degree, with lengthy classifications and polysyllabic terminology. The problem was that geologists were able to recognize distinctive kinds of sedimentary basin fills associated with orogenic belts, and characteristic histories of subsequent deformation of those sediment fills, but no one really knew the tectonic significance of geosynclines. The universal acceptance of plate tectonics has provided a rational framework in which to interpret the development and history of sedimentary basins once called geosynclines. Plate tectonics has simply made the geosyncline concept obsolete.

About the only term in common use that is left over from the heyday of geosynclines is miogeocline, for the prograding wedge of mostly shallow-water sediment at a continental margin. The sediment thickens sharply oceanward and pass into thinner deep-water sediments.

Classification of Sedimentary Basins

Nowadays sedimentary basins are classified by tectonic (and, specifically, plate-tectonic) setting. That's fairly easy to do for modern basins, but it's rather difficult to do for ancient basins. (By modern basins I mean those still within their original tectonic setting; by ancient basins I mean those now separated from their original tectonic setting.) This emphasizes the need for good description and characterization, even if some kind of formal descriptive classification is not actually used. In the following pages is a brief account of the most important kinds of sedimentary basins.

Intracratonic Basins

1. Location and tectonic setting: In anorogenic areas on cratons.

2. Tectonic and sedimentary processes: These basins have no apparent connection with plate tectonics. They are thought to reflect very slow thermal subsidence (for times of the order of a hundred million years) after a heating event under the continental lithosphere. But the reasons for depression below the original crustal level are not understood. Erosion during the thermal uplift seems untenable, as does lithospheric stretching. Was the lithosphere made denser in the area under the basin? Was the lithosphere thinned by "erosion" from beneath? Whatever the reason, the subsequent subsidence can be modeled very well by cooling and isostatic adjustment.

Subsidence is so slow that there seems to have been no depression of the upper surface of the lithosphere, so depositional environments are mostly the same as those in surrounding areas; the succession is just thicker. These successions are also more complete, however—there are fewer and smaller diastems—so at times the basin must have remained under water while surrounding areas were emergent. (A diastem is a brief interruption in sedimentation, with little or no erosion before sedimentation resumes.)

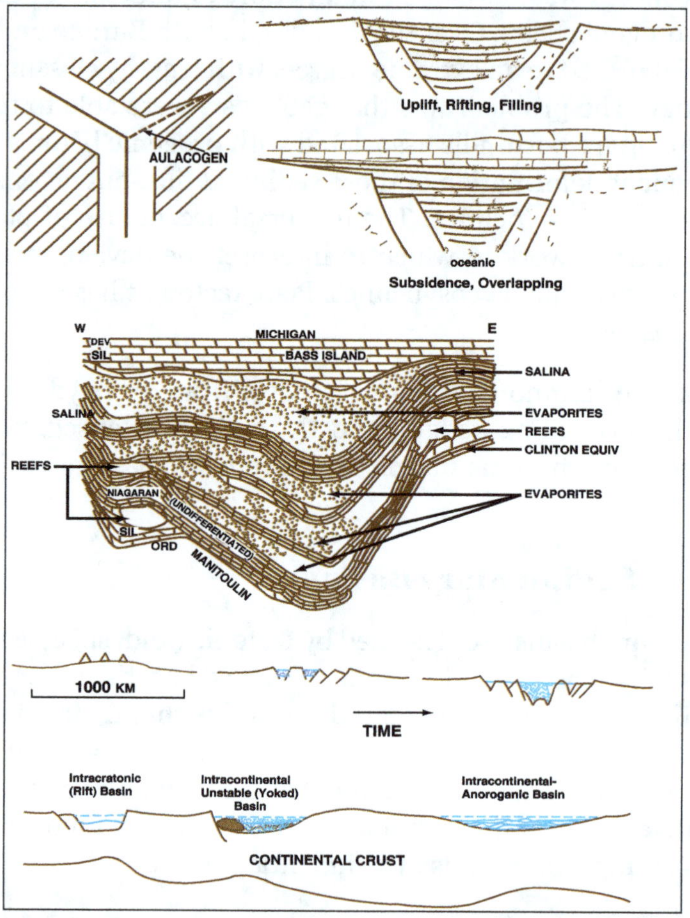

3. Size, shape: Rounded, equidimensional, hundreds of kilometers across.

4. Sediment fill: Shallow-water cratonal sediments (carbonates, shales, sandstones), thicker and more complete than in adjacent areas of the craton but still relatively thin, hundreds of meters.

Aulacogens

1. Location and tectonic setting: extending from the margins toward the interiors of cratons.

2. Tectonic and sedimentary processes: Aulacogens are thought to represent the third, failed arm of a three-armed rift, two of whose arms continued to open to form an ocean basin. In modern settings, aulacogens end at the passive continental margin. An

example is the Benue Trough, underlying the Congo River Basin in West Africa. The ocean eventually closes to form an orogenic belt, so in ancient settings aulacogens end at an orogenic belt; an example is the basin filled by the Pahrump Group (Proterozoic) in Nevada and California.

3. Size, Shape: Long, narrow, linear; tens of kilometers wide, many hundreds of kilometers long.

4. Sediment fill: Very thick (up to several thousand meters); coarse to fine siliciclastics, mostly coarse, minor carbonates; mostly nonmarine, some marine; contemporaneous folding and faulting; the succession often passes upward, with or without major unconformity, into thinner and more widespread shallow-marine cratonal siliciclastics and carbonates.

Rift Basins

1. Location and tectonic setting: Within continental lithosphere on cratons.

2. Tectonic and sedimentary processes: Lithospheric extension on a craton, presumably by regional sublithospheric heating, causes major rifts. Some such rifts continue to open and eventually become ocean basins floored by oceanic rather than continental crust; the basin description here then applies to this earliest stage of the rifting. In other cases, the rifts fail to open fully into ocean basins (perhaps some adjacent and parallel rift becomes the master rift), so they remain floored by thinned continental crust rather than new oceanic crust. A modern example: the East African rift valleys. An ancient example: the Triassic–Jurassic Connecticut and Newark basins in eastern North America.

Sediment supply from the adjacent highlands of the uplifted fault blocks is usually abundant, although in the East African rifts the land slope is away from the rim of the highlands, and surprisingly little sediment reaches the rift basins.

3. Size, Shape: Long, narrow, linear; tens of kilometers wide, up to a few thousand kilometers long.

4. Sediment fill: Coarse to fine siliciclastics, usually nonmarine; often lacustrine sediments; interbedded basalts.

Oceanic Rift Basins

1. Location and tectonic setting: In a narrow and newly opening ocean.

2. Processes: This category of basins is transitional between intracontinental rift basins, described above, and passive-margin basins, described below. Basins described here have opened wide enough to begin to be floored with oceanic crust but are still so narrow that the environment is either still nonmarine or, if marine, has restricted

circulation. Modern examples are the Red Sea and the Gulf of Aden. In the ancient, the sediment fill of such a basin is likely to underlie passive-margin sediments deposited later in the history of ocean opening.

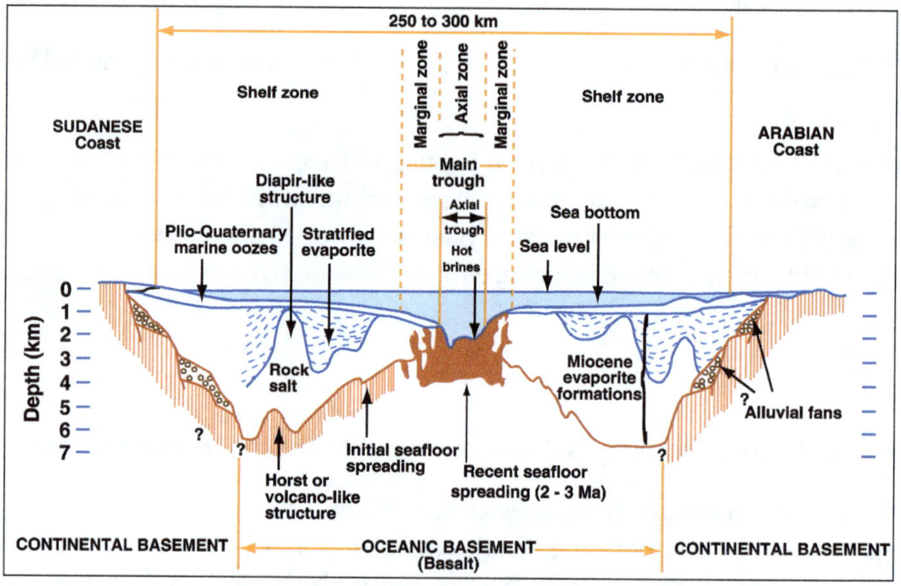

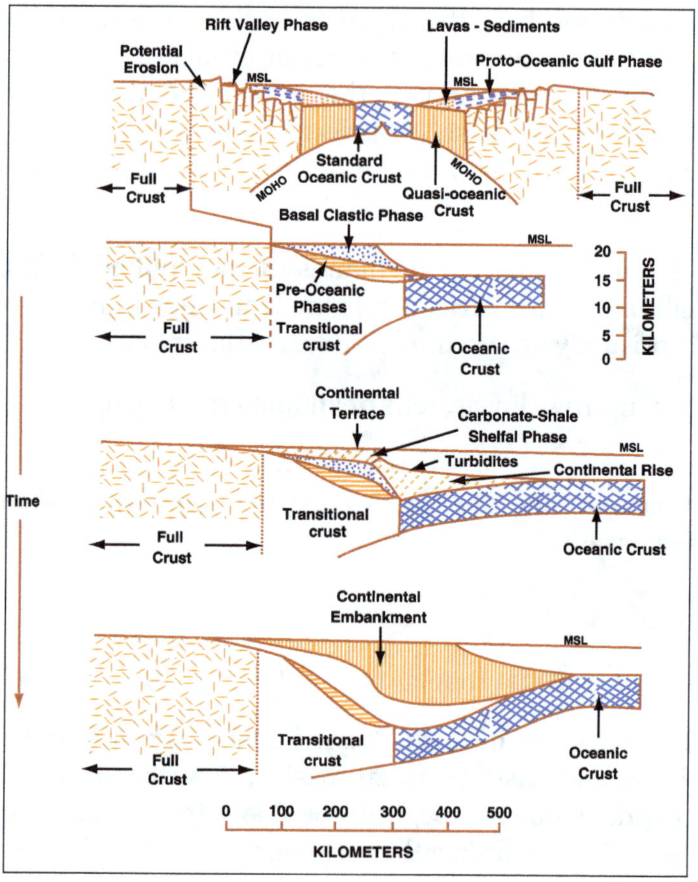

3. Size, Shape: Long, narrow; straight or piecewise straight; tens to a few hundreds of kilometers wide, up to a few thousand kilometers long.

4. Sediment fill: Mafic volcanics and coarse to fine nonmarine siliciclastics, as in intracratonic rift basins described above, passing upward and laterally into evaporites, lacustrine deposits, and fine marine sediments, often metal-rich from hydrothermal activity at the spreading ridge.

Passive Margin Basins

1. Location and tectonic setting: Along passive continental margins, approximately over the transition from continental to oceanic crust formed by rifting and opening of a full-scale ocean basin.

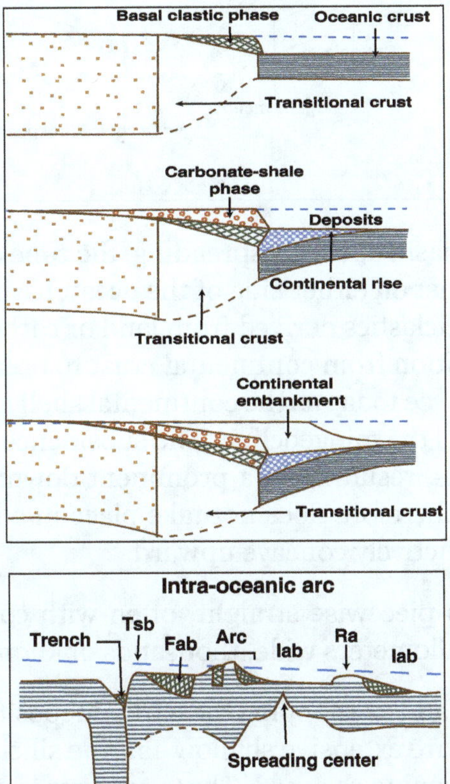

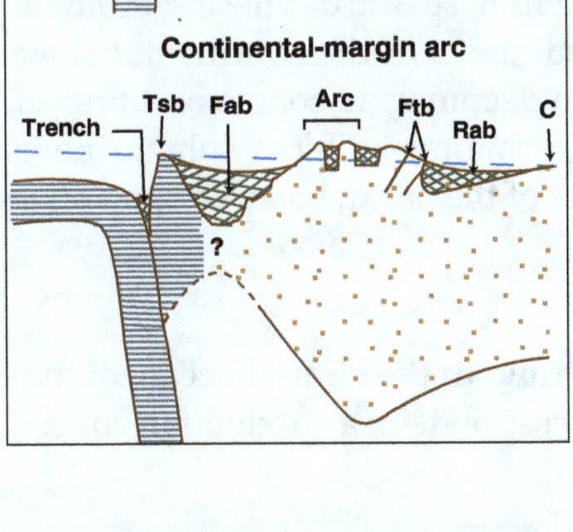

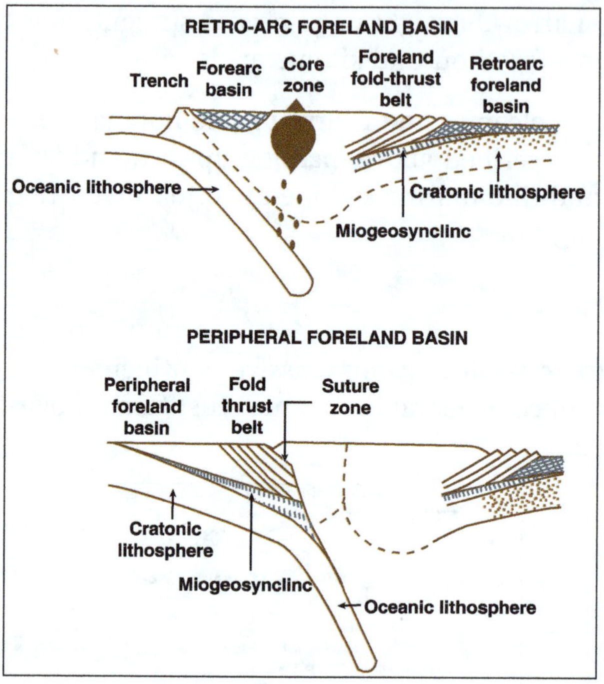

2. Processes: As an ocean basin opens by spreading, the zone of heating and extensional thinning of continental crust on either side of the ocean basin subsides slowly by cooling. Sediments, either siliciclastics derived from land or carbonates generated in place, cover this subsiding transition from continental crust to oceanic crust with a wedge of sediment to build what we see today as the continental shelf and slope. In the context of the ancient, this represents the miogeocline. The subsidence is accentuated by loading of the deposited sediments, resulting in a prominent downbowing of the continental margin. Deposition itself therefore does not take place in a basinal geometry, but the base of the deposit is distinctively concave upward.

3. Size, shape: Straight to piecewise straight, often with considerable irregularity in detail; a few hundreds of kilometers wide, thousands of kilometers long.

4. Sediment fill: Overlying and overlapping the earlier deposits laid down earlier during rifting and initial opening are extensive shallow-marine siliciclastics and carbonates of the continental shelf, thickening seaward. These sediments pass gradually or abruptly into deeper marine fine sediments of the continental slope and rise, often grading or interfingering seaward into deepmarine coarse and fine siliciclastics or resedimented carbonates in the form of turbidites building submarine fans at the base of the slope and filling the deepest parts of the ocean basin to form abyssal plains.

Trenches

1. Location and tectonic setting: In the abyssal ocean, at the line of initial downbending of the subducted oceanic-crust plate in a subduction zone.

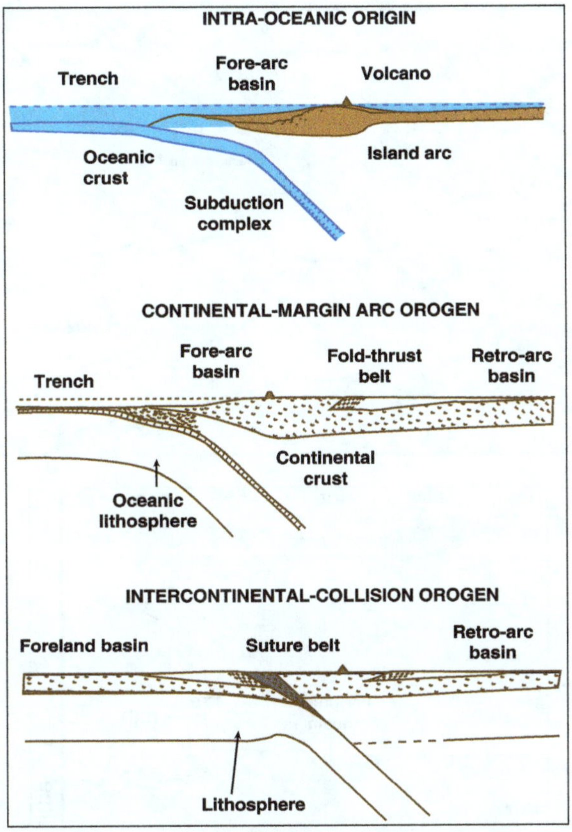

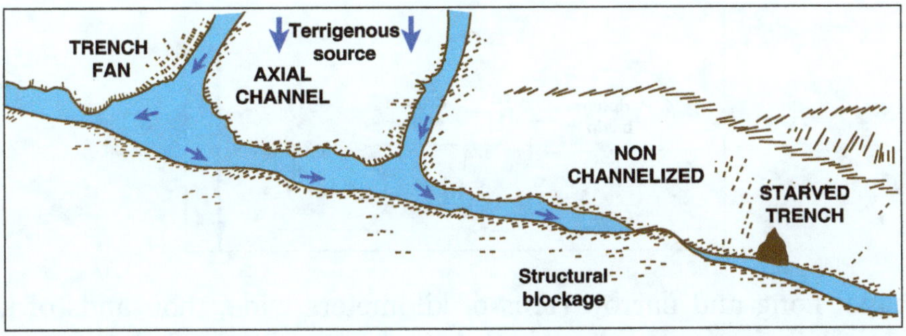

2. Tectonic and sedimentary processes: (1) Open-ocean pelagic sediments (mainly abyssal brown clay and organic oozes) are conveyor-belted to the trench, and underlie the sediments, thin to thick, deposited in the trench itself. (2) What happens to the sediments delivered to or deposited in the trench? While still within the trench they are little deformed, but they don't stay that way long. They are either scraped off the descending plate to form an accretionary wedge, whose structure ranges from chaotically mixed material in a subduction mélange to a fairly regular imbricated succession of underthrusted sheets dipping toward the arc (note: the thrust sheets themselves get younger downward, but within a given plate the sediments get younger upward), or they are dragged down the subduction zone. The extent of the latter process is controversial, but cases have been described.

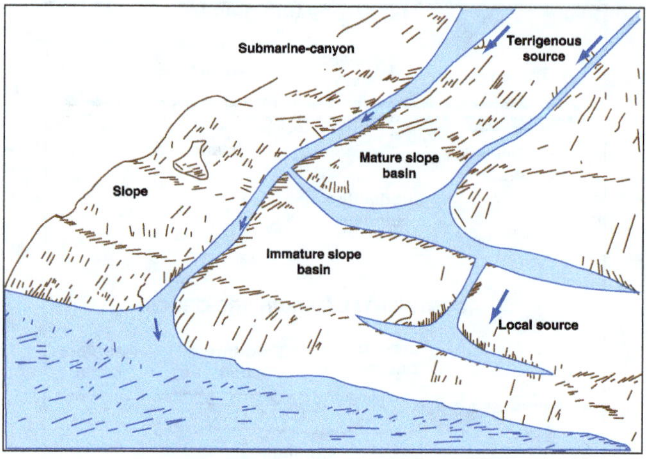

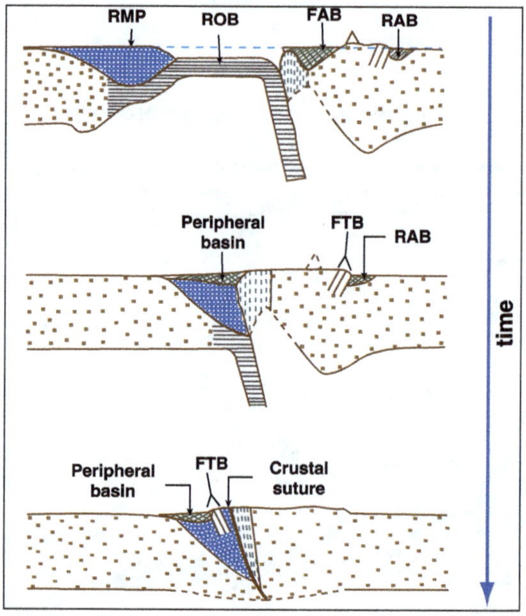

3. Size, Shape: Long and narrow (tens of kilometers wide, thousands of kilometers long), arcuate, with convex side toward the oncoming subducted plate.

4. Sediment fill: Varies from thin (hundreds of meters) pelagic sediments (fine abyssal muds, volcanic ash) to thick (thousands of meters) arc-derived coarse siliciclastics and volcaniclastics, as local fans built perpendicular to the trench axis or oblong fanlike bodies built parallel to trench axis.

Trench-Slope Basins

1. Location and tectonic setting: On the inner (arcward) wall of subduction-zone trenches.

2. Tectonic and sedimentary processes: The basins are formed as low areas, with closed contours, between adjacent thrust sheets in the growing accretionary wedge.

Near-surface folding of sediment in the accretionary wedge may also be a factor in the development of the basins. These basins intercept some of the sediment carried as turbidity currents from upraised older parts of the accretionary complex, or from the more distant arc itself.

3. Size, shape: Small (no larger than kilometers wide, tens of kilometers long, often smaller); linear, and elongated parallel to the trench.

4. Sediment fill: Deep-marine silts and muds sedimented directly into the basins or slumped into the basins from higher on the slope; also coarser siliciclastics supplied from farther upslope by turbidity currents.

Fore-Arc Basins

1. Location and tectonic setting: In subduction zones; between the upraised subduction complex just inboard of the trench and the volcanic arc (in the case of ocean-ocean subduction) or the overriding continent (in the case of ocean–continent subduction).

2. Processes: As subduction proceeds, a relatively low area, usually below sea level, is formed between the relatively high outer arc upraised by subduction and the inner volcanic arc built by subduction magmatism. Ancient examples of such fore-arc basins are likely to be tectonically isolated from the originally adjacent areas. After an arc–continent collision, another variant of fore-arc basin can be formed between the outer arc and the overriding continent. These basins are likely to be filled mainly from the high land of the tectonically active continent. Later continent–continent collision would make direct reconstruction of their tectonic setting difficult.

3. Size, shape: Tens of kilometers to over one hundred kilometers wide, up to thousands of kilometers long; commonly arcuate.

4. Sediment fill: Nonmarine siliciclastic fluvial to deltaic deposits at the arcward margin pass seaward into deep marine siliciclastics, mainly sediment-gravityflow deposits, all interbedded with arc-derived volcanics flows and pyroclastics. Section thickness can be many thousands of meters.

Foreland Basins

1. Location, tectonic setting, processes: There are two kinds of foreland basins: retro-arc foreland basins, which are formed on stable continental crust by loading by thrust sheets moving toward the continental interior as a result of compression and crustal shortening in an ocean–continent subduction zone, and peripheral foreland basins, formed after continent–continent collisions by loading of the continental crust of the subducted plate by development of thrust sheets in the continental crust of the subducted plate directed back away from the subduction zone. Both kinds tend to be asymmetrical, with their deepest parts nearest the emplaced thrust sheets. They tend

to migrate away from the arc or suture zone with time. They are filled by sediments derived from the mountainous terrain associated with the compression and thrusting.

2. Size, Shape: Tens to a few hundreds of kilometers wide, hundreds to thousands of kilometers long; often with varying development along their length; commonly arcuate or piecewise arcuate, reflecting the geometry of subduction.

3. Sediment fill: Coarse fluvial siliciclastics, mainly as alluvial fans, thinning and fining away from the arc or suture, often passing into shallow-marine sandstone–shale successions if sea level is high enough to flood the basin. Thicknesses are up to many thousands of meters. The classic molasse facies, thick nonmarine conglomerates, is deposited in foreland basins.

Remnant Basins

1. Location and tectonic setting: Within suture zones formed by continent–continent collision.

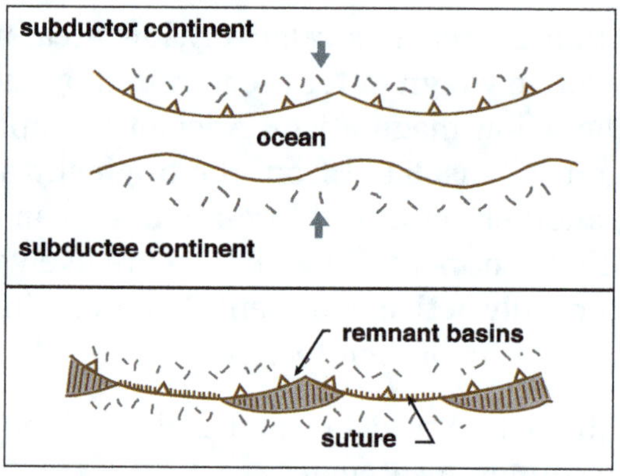

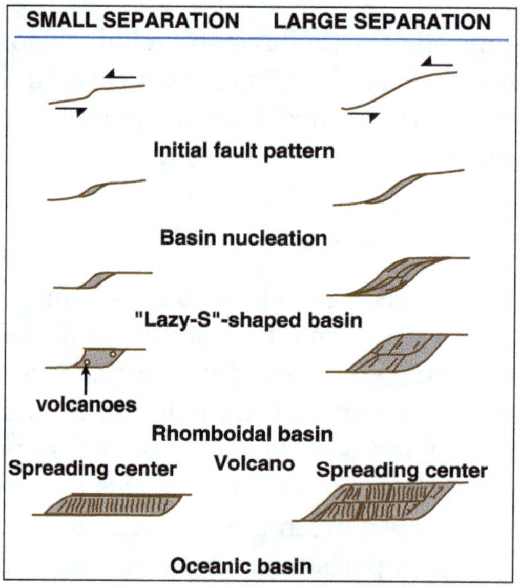

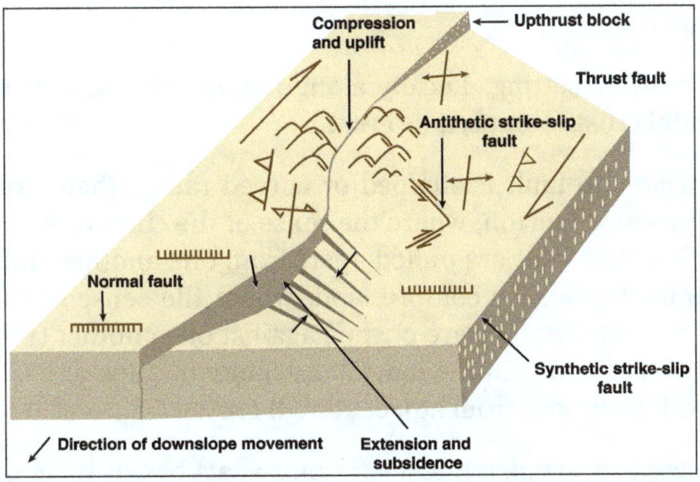

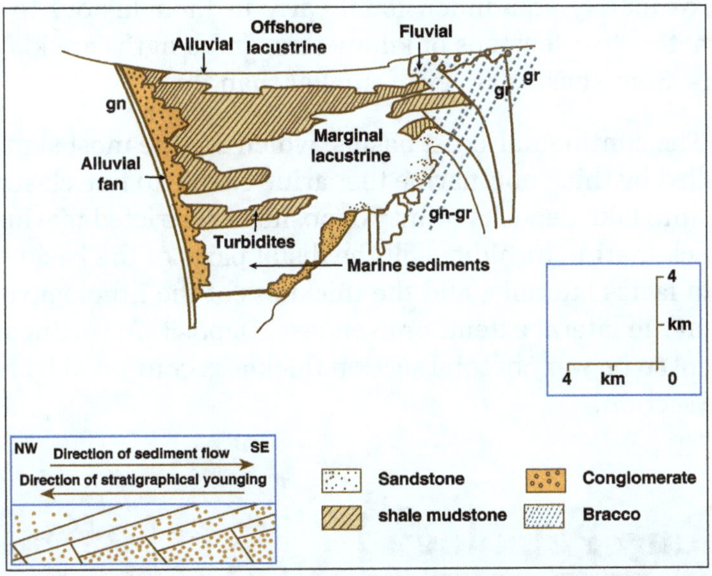

2. Processes: Continental margins and subduction zones are (for different reasons, connected with geometry of rifting and geometry of subduction) commonly irregular in plan rather than straight, so when continent–continent collision eventually comes to pass, certain salients of continental crust encounter the subduction zone before reentrants. With further subduction and suturing, this creates isolated basins still floored by residual oceanic crust, which receive abundant sediment from adjacent strongly uplifted crust.

3. Size, shape: Many tens to hundreds of kilometers across; irregular in shape.

4. Sediment fill: Very thick and highly varied, with strong lateral facies changes; usually fluvial at the margins, commonly passing into deep-marine sedimentgravity-flow deposits; sometimes the basin becomes sealed off from the ocean, so that facies include lacustrine sediments.

Pull-Apart Basins

1. Location and tectonic setting: Locally along major strike-slip or transform faults, either in continental crust or in oceanic crust.

Processes: If a strike-slip fault is stepped or curved rather than straight, movement along it tends to produce tension, where the sense of the curvature and movement are such that the walls of the fault are pulled apart from one another (this kind of regime is described as transtensile), or compression, where the sense of the curvature and movement are such that the walls are pushed against one another (this kind of regime is described as transpressive). In the tensional segments, gaps or basins are produced which are filled with sediment from adjacent high crust.

3. Size, shape: There is a strong tendency for pull-apart basins to be rhomboidal. They range from approximately equidimensional early in their history to elongated later. Widths are kilometers to a few tens of kilometers, and lengths are kilometers to many tens of kilometers. Some basins are even smaller than this.

4. Sediment fill: The continental-crust basins, which are the most significant sedimentologically, are filled by thick nonmarine to marine coarse to fine clastics, often as alluvial fans passing into lake deposits or into deposits of restricted marine environments. In some cases thick marine turbidites fill the distal parts of the basin. There is usually sharp variation in facies laterally, and the thickness of the lithologic units may be not much greater than the lateral extent, or even less. Deposition is concurrent with elongation of the basin, so be wary of total section thickness computed by bed-by-bed measurements of the section.

Sedimentary Petrology

The field of sedimentary petrology is concerned with the description and classification of sedimentary rocks, interpretation of the processes of transportation and deposition of the sedimentary materials forming the rocks, the environment that prevailed at the time the sediments were deposited, and the alteration (compaction, cementation, and chemical and mineralogical modification) of the sediments after deposition.

There are two main branches of sedimentary petrology. One branch deals with carbonate rocks, namely limestones and dolomites, composed principally of calcium carbonate (calcite) and calcium magnesium carbonate (dolomite). Much of the complexity in classifying carbonate rocks stems partly from the fact that many limestones and dolomites have been formed, directly or indirectly, through the influence of organisms, including bacteria, lime-secreting algae, various shelled organisms (e.g., mollusks and brachiopods), and by corals. In limestones and dolomites that were deposited under marine conditions, commonly in shallow warm seas, much of the material initially forming the rock consists of skeletons of lime-secreting organisms. In many examples, this skeletal material is preserved as fossils. Some of the major problems of carbonate petrology concern the physical and biological conditions of the environments in which carbonate material has been deposited, including water depth, temperature, degree of illumination by sunlight, motion by waves and currents, and the salinity and other chemical aspects of the water in which deposition occurred.

Banded-iron formation (BIF): A banded-iron formation (BIF) rock recovered from the Temagami greenstone belt in Ontario, Canada, and dated to 2.7 billion years ago. Dark layers of iron oxide are intercalated with red chert.

The other principal branch of sedimentary petrology is concerned with the sediments and sedimentary rocks that are essentially noncalcareous. These include sands and sandstones, clays and claystones, siltstones, conglomerates, glacial till, and varieties of sandstones, siltstones, and conglomerates (e.g., the graywacke-type sandstones and siltstones). These rocks are broadly known as clastic rocks because they consist of distinct particles or clasts. Clastic petrology is concerned with classification, particularly with respect to the mineral composition of fragments or particles, as well as the shapes of particles (angular versus rounded), and the degree of homogeneity of particle sizes. Other main concerns of clastic petrology are the mode of transportation of sedimentary materials, including the transportation of clay, silt, and fine sand by wind; and the transportation of these and coarser materials through suspension in water, through traction by waves and currents in rivers, lakes, and seas, and sediment transport by ice.

Sedimentary petrology also is concerned with the small-scale structural features of sediments and sedimentary rocks. Features that can be conveniently seen in a specimen held in the hand are within the domain of sedimentary petrology. These features include the geometrical attitude of mineral grains with respect to each other, small-scale cross stratification, the shapes and interconnections of pore spaces, and the presence of fractures and veinlets.

Instruments and methods used by sedimentary petrologists include the petrographic microscope for description and classification, X-ray mineralogy for defining fabrics and small-scale structures, physical model flume experiments for studying the effects of flow as an agent of transport and the development of sedimentary structures, and mass spectrometry for calculating stable isotopes and the temperatures of deposition, cementation, and diagenesis. Wet-suit diving permits direct observation of current processes on coral reefs, and manned submersibles enable observation at depth on the ocean floor and in mid-oceanic ridges.

The plate-tectonic theory has given rise to much interest in the relationships between sedimentation and tectonics, particularly in modern plate-tectonic environments—e.g., spreading-related settings (intracontinental rifts, early stages of intercontinental rifting such as the Red Sea, and late stages of intercontinental rifting such as the margins of the present Atlantic Ocean), mid-oceanic settings (ridges and transform faults), subduction-related settings (volcanic arcs, fore-arcs, back-arcs, and trenches), and continental collision-related settings (the Alpine-Himalayan belt and late orogenic basins with molasse [i.e., thick association of clastic sedimentary rocks consisting chiefly of sandstones and shales]). Today many subdisciplines of sedimentary petrology are concerned with the detailed investigation of the various sedimentary processes that occur within these plate-tectonic environments.

References

- Sedimentology: spe.org, Retrieved 21, July 2020
- Depoenvirons: wvc.edu, Retrieved 09, March 2020
- Sedimentary-petrology, geology, science: britannica.com, Retrieved 17, August 2020

Chapter 2

Sediments and Sedimentary Rocks

The naturally occurring materials which are broken down through the processes such as erosion and weathering are known as sediments. Some of the different types of sediments are biogenic sediments and marine sediments. The topics elaborated in this chapter will help in gaining a better perspective about these types of sediments as well as sedimentary rocks.

Sediment is solid material that is moved and deposited in a new location. Sediment can consist of rocks and minerals, as well as the remains of plants and animals. It can be as small as a grain of sand or as large as a boulder.

Sediment moves from one place to another through the process of erosion. Erosion is the removal and transportation of rock or soil. Erosion can move sediment through water, ice, or wind.

Water can wash sediment, such as gravel or pebbles, down from a creek, into a river, and eventually to that river's delta. Deltas, river banks, and the bottom of waterfalls are common areas where sediment accumulates.

Glaciers can freeze sediment and then deposit it elsewhere as the ice carves its way through the landscape or melts. Sediment created and deposited by glaciers is called moraine.

Wind can move dirt across a plain in dust storms or sandstorms. Sand dunes are made of rocky sediment worn down by wind and collision with other sand particles.

Sediment is important because it often enriches the soil with nutrients. Areas rich in sediments are often also rich in biodiversity. Sedimentary soil is usually better for farming. Deltas and river banks, where much sediment is deposited, are often the most fertile agricultural areas in a region.

Sedimentary Rocks

Rivers, oceans, winds, and rain runoff all have the ability to carry the particles washed off of eroding rocks. Such material, called detritus, consists of fragments of rocks and minerals. When the energy of the transporting current is not strong enough to carry these particles, the particles drop out in the process of sedimentation. This type of sedimentary deposition is referred to as clastic sedimentation. Another type of sedimentary deposition occurs when material is dissolved in water, and chemically precipitates from the water. This type of sedimentation is referred to as chemical sedimentation. A third process can occur, wherein living organisms extract ions dissolved in water to make such things as shells and bones. This type of sedimentation is called biochemical sedimentation. The accumulation of plant matter, such as at the bottom of a swamp, is referred to as organic sedimentation. Thus, there are 4 major types of sedimentary rocks: Clastic Sedimentary Rocks, Chemical Sedimentary Rocks, Biochemical Sedimentary Rocks, and Organic Sedimentary Rocks.

Clastic Sediments and Sedimentary Rocks

The formation of clastic sediment and sedimentary rocks involves five processes:

- Weathering: The first step is transforming solid rock into smaller fragments or dissolved ions by physical and chemical weathering as discussed in the last lecture.
- Erosion: Erosion is actually many processes which act together to lower the surface of the earth. In terms of producing sediment, erosion begins the transportation process by moving the weathered products from their original location. This can take place by gravity (mass movement events like landslides or rock falls), by running water, by wind, or by moving ice. Erosion overlaps with transportation.
- Transportation: Sediment can be transported by sliding down slopes, being picked up by the wind, or by being carried by running water in streams, rivers, or ocean currents. The distance the sediment is transported and the energy of the transporting medium all leave clues in the final sediment that tell us something about the mode of transportation.
- Deposition: Sediment is deposited when the energy of the transporting medium becomes too low to continue the transport process. In other words, if the velocity of the transporting medium becomes too low to transport sediment, the sediment will fall out and become deposited. The final sediment thus reflects the energy of the transporting medium.
- Lithification (Diagenesis): Lithification is the process that turns sediment into rock. The first stage of the process is compaction. Compaction occurs as the weight of the overlying material increases. Compaction forces the grains closer together, reducing pore space and eliminating some of the contained water. Some of this water may carry mineral components in solution, and these constituents may later precipitate as new minerals in the pore spaces. This causes cementation, which will then start to bind the individual particles together.

Classification: Clastic sedimentary particles and sedimentary rocks are classified in terms of grain size and shape, among other factors.

Name of Particle	Size Range	Loose Sediment	Consolidated Rock
Boulder	>256 mm	Gravel	Conglomerate or Breccia (depends on rounding)
Cobble	64 - 256 mm	Gravel	
Pebble	2 - 64 mm	Gravel	
Sand	1/16 - 2mm	Sand	Sandstone
Silt	1/256 - 1/16 mm	Silt	Siltstone
Clay	<1/256 mm	Clay	Claystone, mudstone, and shale

In general, the coarser sediment gets left behind by the transportation process. Thus, coarse sediment is usually found closer to its source and fine grained sediment is found farther from the source.

Textures of Clastic Sedimentary Rocks

When sediment is transported and deposited, it leaves clues to the mode of transport and deposition. For example, if the mode of transport is by sliding down a slope, the deposits that result are generally chaotic in nature, and show a wide variety of particle sizes. Grain size and the interrelationship between grains give the resulting sediment texture. Thus, we can use the texture of the resulting deposits to give us clues to the mode of transport and deposition.

1. Sorting - The degree of uniformity of grain size. Particles become sorted on the basis of density, because of the energy of the transporting medium. High energy currents can carry larger fragments. As the energy decreases, heavier particles are deposited and lighter fragments continue to be transported. This results in sorting due to density.

If the particles have the same density, then the heavier particles will also be larger, so the sorting will take place on the basis of size. We can classify this size sorting on a relative basis - well sorted to poorly sorted. Sorting gives clues to the energy conditions of the transporting medium from which the sediment was deposited.

Poorly Sorted Sediment Well Sorted Sediment

Examples:

- Beach deposits and windblown deposits generally show good sorting because the energy of the transporting medium is usually constant.
- Stream deposits are usually poorly sorted because the energy (velocity) in a stream varies with position in the stream and time.

2. Rounding - During the transportation process, grains may be reduced in size due to abrasion. Random abrasion results in the eventual rounding off of the sharp corners and edges of grains. Thus, rounding of grains gives us clues to the amount of time sediment has been in the transportation cycle. Rounding is classified on relative terms as well.

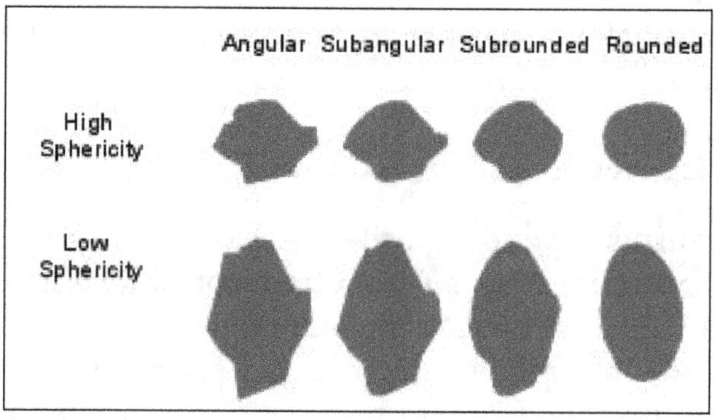

Sediment Maturity

Sediment Maturity refers to the length of time that the sediment has been in the sedimentary cycle. Texturally mature sediment is sediment that is well rounded, (as rounding increases with transport distance and time) and well sorted (as sorting gets better as larger clasts are left behind and smaller clasts are carried away. Because the weathering processes continues during sediment transport, mineral grains that are unstable near the surface become less common as the distance of transport or time in the cycle increases. Thus compositionally mature sediment is composed of only the most stable minerals.

For example a poorly sediment containing glassy angular volcanic fragments, olivine crystals and plagioclase is texturally immature because the fragments are angular, indicating they have not been transported very far and the sediment is poorly sorted, indicating that little time has been involved in separating larger fragments from smaller fragments. It is compositionally immature because it contains unstable glass along with minerals that are not very stable near the surface - olivine and plagioclase.

On the other hand well sorted beach sand consisting mainly of well-rounded quartz grains is texturally mature because the grains are rounded, indicating a long time in the transportation cycle, and the sediment is well sorted, also indicative of the long time required to separate the coarser grained material and finer grained material from the sand. The beach sand is compositionally mature because it is made up only of quartz which is very stable at the earth's surface.

Types of Clastic Sedimentary Rocks

Conglomerates and Breccias

Conglomerate and Breccia are rocks that contain an abundance of coarse grained clasts (pebbles, cobbles, or boulders). In a conglomerate, the coarse grained clasts are well rounded, indicating that they spent considerable time in the transportation process and were ultimately deposited in a high energy environment capable of carrying the large clasts. In a breccia, the coarse grained clasts are very angular, indicating the clasts spent little time in the transportation cycle.

Sandstones

Sandstone is made of sand-sized particles and forms in many different depositional settings. Texture and composition permit historic interpretation of the transport and depositional cycle and sometimes allows determination of the source. Quartz is, by far, the dominant mineral in sandstones. Still there are other varieties. A Quartz arenite – is nearly 100% quartz grains. An Arkose contains abundant feldspar. In lithic sandstone, the grains are mostly small rock fragments. A Wacke is a sandstone that contains more than 15% mud (silt and clay sized grains). Sandstones are one of the most common types of sedimentary rocks.

Mudrocks

Mudrocks are made of fine grained clasts (silt and clay sized). A siltstone is one variety that consists of silt-sized fragments. Shale is composed of clay sized particles and is a rock that tends to break into thin flat fragments. A mudstone is similar to shale, but does not break into thin flat fragments. Organic-rich shales are the source of petroleum.

Fine grained clastics are deposited in non-agitated water, calm water, where there is little energy to continue to transport the small grains. Thus mudrocks form in deep water ocean basins and lakes.

Biochemical and Organic Sediments and Sedimentary Rocks

Biochemical and Organic sediments and sedimentary rocks are those derived from living organisms. When the organism dies, the remains can accumulate to become sediment or sedimentary rock. Among the types of rock produced by this process are:

- Biochemical Limestone - Calcite ($CaCO_3$) is precipitated by organisms usually to form a shell or other skeletal structure. Accumulation of these skeletal remains results in a limestone. Sometimes the fossilized remains of the organism are preserved in the rock, other times recrystallization during lithification has destroyed the remains. Limestones are very common sedimentary rocks.

- Biochemical Chert - Tiny silica secreting planktonic organism like Radiolaria and Diatoms can accumulate on the sea floor and recrystallize during lithification to form biochemical chert. The recrystallization results in a hard rock that is usually seen as thin beds.
- Diatomite - When diatoms accumulate and do not undergo recrystallization, they form a white rock called diatomite as seen in the White Cliffs of Dover.
- Coal - Coal is an organic rock made from organic carbon that is the remains of fossil plant matter. It accumulates in lush tropical wetland settings and requires deposition in absence of Oxygen. It is high in carbon and can easily be burned to obtain energy.

Chemical Sediments and Sedimentary Rocks

Dissolved ions released into water by the weathering process are carried in streams or groundwater. Eventually these dissolved ions end in up in the ocean, explaining why sea water is salty. When water evaporates or the concentration of the ions get too high as a result of some other process, the ions recombine by chemical precipitation to form minerals that can accumulate to become chemical sediments and chemical sedimentary rocks. Among these are:

- Evaporites - Formed by evaporation of sea water or lake water. Produces halite (salt) and gypsum deposits by chemical precipitation as concentration of solids increases due to water loss by evaporation. This can occur in lakes that have no outlets (like the Great Salt Lake) or restricted ocean basins, like has happened in the Mediterranean Sea or the Gulf of Mexico in the past.
- Travertine - Groundwater containing dissolve Calcium and bicarbonate ions can precipitate calcite to form a chemically precipitated limestone, called travertine. This can occur in lakes, hot springs, and caves.
- Dolostones - Limestone that have been chemically modified by Mg-rich fluids flowing through the rock are converted to dolostones. $CaCO_3$ is recrystallized to a new mineral dolomite $CaMg(CO_3)_2$.
- Chemical Cherts - Groundwater flowing through rock can precipitate SiO_2 to replace minerals that were present. This produces a non-biogenic chert. There are many varsities of such chert that are given different names depending on their attributes, For example:
    - Flint – Black or gray from organic matter.

- Jasper – Red or yellow from Fe oxides.
- Petrified wood – Wood grain preserved by silica.
- Agate – Concentrically layered rings.

Biogenic Sediments

Biogenic sediments, which are defined as containing at least 30% skeletal remains of marine organisms, cover approximately 62% of the deep ocean floor. Clay minerals make up most of the non-biogenic constituents of these sediments. While a vast array of plants and animals contribute to the organic matter that accumulates in marine sediments, a relatively limited group of organisms contribute significantly to the production of biogenic deep-sea sediments, which are either calcareous or siliceous oozes.

Distributions and accumulation rates of biogenic oozes in oceanic sediments depend on three major factors:

- Rates of production of biogenic particles in the surface waters.
- Dissolution rates of those particles in the water column and after they reach the bottom.
- Rates of dilution by terrigenous sediments.

The abundances and distributions of the organisms that produce biogenic sediments depend upon such environmental factors as nutrient supplies and temperature in the oceanic waters in which the organisms live. Dissolution rates are dependent upon the chemistry of the deep ocean waters through which the skeletal remains settle and of the bottom and interstitial waters in contact with the remains as they accumulate and are buried. The chemistry of deep-sea waters, is, in turn, influenced by the rate of supply of both skeletal and organic remains of organisms from surface waters. It is also heavily dependent upon the rates of deep ocean circulation and the length of time that the bottom water has been accumulating CO_2 and other byproducts of biotic activities.

Carbonate Oozes

Production

Most carbonate or calcareous oozes are produced by the two different groups of organisms. The major constituents of nanofossil or coccolith ooze are tiny (less than 10 microns) calcareous plates produced by phytoplankton of the marine algal group, the Coccolithophoridae or by an extinct group called discoasters. Foraminiferal ooze is dominated by the tests (shells) of planktic protists belonging to the Foraminiferida.

Most foraminiferal tests are sand-sized (>61 mm in diameter), so many foraminiferal oozes are bimodal in particle-size distribution, because they are made up of sand-sized foraminiferal tests and mud-sized coccolith plates.

Discoasters, coccoliths and foraminiferal tests are all made of the mineral calcite. Pteropod ooze is produced by the accumulation of shells of pteropods and heteropods, which are small planktic mollusks. As these shells are composed of the mineral aragonite, pteropod oozes are more easily dissolved, so are restricted to relatively shallow depths (less than 3,000 m) in tropical areas.

Carbonate oozes are the most widespread shell deposits on earth. Nearly half the pelagic sediment in the world's oceans is carbonate ooze. Furthermore, foraminifera and coccolithophorids have been major producers of pelagic sediment for the past 200 million years. As a result, these are arguably among the most important and scientifically useful organisms on Earth. Because their larger size makes them easier to identify and work with, this is particularly true for the foraminifera. Their fossils provide the single most important record of Earth history over the past 200 million years. That history is recorded not only by the evolution of species and higher taxa through that time, but is also preserved in the chemistry of the fossils themselves. The field of Paleoceanography owns much of its existence to biostratigraphy, isotope stratigraphy and paleoenvironmental analyses that utilize fossil foraminifera.

The distributions and abundances of living planktic foraminifera and coccolithophorids in the upper few hundred meters of the ocean depends in large part on nutrient supply and temperature. Coccolithophorids, because they are marine algae, require sunlight and inorganic nutrients (fixed N, P, and trace nutrients) for growth. However, most coccolithophorid species grow well with very limited supplies of nutrients and do not compete effectively with diatoms and dinoflagellates when nutrients are plentiful. Furthermore, both high nutrient supplies and cold temperatures inhibit calcium carbonate production to some degree. For these reasons, diversities (number of different kinds) of coccolithophorids are high and production rates of coccoliths are moderate even in the most nutrient-poor regions of the subtropical oceans, the subtropical gyres. Production of coccoliths is higher in equatorial upwelling zones and often along continental margins and in temperate latitudes where nutrient supplies are higher, though diversities decline. In very high nutrient areas, such as upwelling zones in the eastern tropical oceans (*i.e.*, meridional upwelling), polar divergences and near river mouths, production of coccoliths is minimal.

Even though planktic foraminifera are protozoans rather than algae, their distributions, diversities, and carbonate productivity are quite similar to those of coccolithophorids. Many planktic foraminifera, especially the spinose species that live in the upper 100 m of temperate to tropical oceans host dinoflagellate symbionts which aid the foraminifera by providing energy and enhancing calcification. Having algal symbionts is highly advantageous in oceanic waters where inorganic nutrients and food are scarce, so a diverse assemblage of planktic foraminifera thrives along with the coccolithophorids in the nutrient-poor subtropical gyres. Greater abundances of fewer species thrive in equatorial upwelling zones and along continental margins, so rates of carbonate shell production are higher. And similar to coccolithophorids, few planktic foraminifera live in very high nutrient areas, such as upwelling zones in the eastern tropical oceans, polar divergences and near river mouths, so production of carbonate sediments is minimal in these areas. Finally, planktic foraminifera require deep oceanic waters to complete their life cycles, which they cannot do in neretic waters over continental shelves.

Cool temperatures work together with higher nutrient supplies to reduce diversities of coccolithophorids and planktic foraminifera, and ultimately to shift the ecological community to organisms that do not produce carbonate sediments. A 10 °C drop in temperature is physiologically similar to doubling nutrient supply, which is why the pelagic community in an equatorial upwelling zone resembles that of a temperate oceanic region, while the pelagic community of an intensive meridional upwelling zone resembles subpolar to polar communities.

If surface production was the only factor controlling accumulation rates of carbonate oozes, deep-sea sediment patterns would be quite simple. Carbonate oozes would cover the seafloor everywhere except.

- Beneath intensive meridional upwelling zones.
- Beneath polar seas.
- Where they are overwhelmed by terrigenous sedimentation.

Rates of accumulation would be on the order of 3-5 cm/1000 years in the Open Ocean and 10-20 cm/year beneath equatorial upwelling zones and along most continental margins.

Dissolution

Over much of the ocean floor, carbonate accumulation rates are controlled more by dissolution in bottom waters than by production in surface waters. Dissolution of calcium carbonate in seawater is influenced by three major factors: temperature, pressure and partial pressure of carbon dioxide (CO_2). The easiest way to understand calcium carbonate ($CaCO_3$) dissolution is to recognize that it is controlled, in large part, by the solubility of CO_2:

$$CaCO_3 + H_2O + CO_2 \Leftrightarrow Ca^{++} + 2HCO_3^-$$

The more CO_2 that can be held in solution, the more $CaCO_3$ that will dissolve. Since more CO_2 can be held in solution at higher pressures and cooler temperatures, $CaCO_3$ is more soluble in the deep ocean than in surface waters. Finally, as CO_2 is added to the water, more $CaCO_3$ can dissolve. The result is that, as more CO_2 is added to deep ocean water by the respiration of organisms, the more corrosive the bottom water becomes to calcareous shells.

The rain of organic matter from surface waters through time increases the partial pressure of CO_2 in bottom water, so the longer the bottom water has been out of contact with the surface, the higher its partial pressure of CO_2. Beneath high-nutrient surface waters, primary production exceeds what is utilized in the surface mixed layer. Excess organic matter falling through the water column accumulates on the bottom, where organisms feed upon it and oxidize it to CO_2.

The depth at which surface production of $CaCO_3$ equals dissolution is called the calcium carbonate compensation depth (CCD). Above this depth, carbonate oozes can accumulate, below the CCD only terrigenous sediments, oceanic clays, or siliceous oozes can accumulate. The calcium carbonate compensation depth beneath the temperate and tropical Atlantic is approximately 5,000 m deep, while in the Pacific, it is shallower, about 4,200-4,500 m, except beneath the equatorial upwelling zone, where the CCD is about 5,000 m. The CCD in the Indian Ocean is intermediate between the Atlantic and the Pacific. The CCD is relatively shallow in high latitudes.

Surface waters of the ocean tend to be saturated with respect to $CaCO_3$; low latitude surface waters are usually supersaturated. At shallow to intermediate seafloor depths (less than 3000 m), foraminiferal tests and coccolith plates tend to be well preserved in bottom sediments. However, at depths approaching the CCD, preservation declines as smaller and more fragile foraminiferal tests show signs of dissolution. The boundary zone between well preserved and poorly preserved foraminiferal assemblages is known as the lysocline.

The preservation potential of the various kinds of carbonate shells and skeletons differs. Pteropod shells are aragonite, a less stable form of $CaCO_3$. Pteropod shells dissolve at depths greater than 3,000 m in the Atlantic Ocean and below a few hundred meters in the Pacific. Calcitic planktic foraminiferal tests, especially small tests of juvenile spinose foraminifera, dissolve more readily than coccoliths, which are also made of calcite. Pelagic sediments from relatively shallow depths in low latitudes are often dominated by pteropods shells, at intermediate depths by foraminiferal tests, below the lysocline and above the CCD by coccoliths, and below the CCD by red clays.

Regional changes in the depths of the lysocline and CCD result, in part, from changes in CO_2 content of bottom waters as they "age". In modern oceans, deep ocean circulation is driven by formation of bottom waters during the freezing of sea ice. Seawater, due to its salt content, can cool below -1° C before ice begins to form. When sea ice forms, the salt is excluded and is left behind in the seawater. Water in the vicinity of the freezing sea ice becomes more saline and therefore denser. As a result, large-scale sea ice formation creates very dense water masses that sink to the bottom of the ocean to form deep bottom water. During the Antarctic winter, the freezing of sea ice in the Weddell Sea produces Antarctic Bottom Water (AABW), which sinks to the sea bottom and spreads northward into the South Atlantic. During the Arctic winter, sea ice formation in the Norwegian and Greenland Seas produce North Atlantic Deep Water (NADW), which sinks to the bottom of the North Atlantic and flows southward. AABW is slightly denser than NADW, so when they meet, AABW flows beneath NADW. As the NADW and AABW spread eastward into the Indian and Pacific Oceans, they mix to become Deep Pacific Common Water (DPCW). The "youngest" bottom waters are in the Atlantic, the "oldest" are in the North Pacific.

When seawater is at the surface, it equilibrates with the atmosphere with respect to O_2 and CO_2. From the time a water mass sinks from the surface until it comes back to the surface, respiration by organisms in the water column and on the bottom use up O_2 and add CO_2. As a result, the longer bottom water is away from the surface, the more corrosive it is to $CaCO_3$.

Carbonate Sedimentation Worldwide

The depth of the CCD and the pattern of carbonate sedimentation in any part of the world's ocean reflect the influences of surface production of organic matter, surface production of carbonates, and the corrosiveness of the bottom water to $CaCO_3$.

Because coccolithophorids and planktic foraminifera thrive in temperate to subtropical oceans where surface nutrient supplies are very limited, these organisms produce a continual rain of $CaCO_3$ to the sea floor. In equatorial upwelling zones, organic productivity is elevated enough to stimulate higher rates of production of calcareous and siliceous skeletal remains, but not enough to export excess organic matter to the deep ocean where its respiration would increase corrosiveness of bottom waters to $CaCO_3$.

In more intensive upwelling zones, especially in the eastern tropical Pacific and the Antarctic divergence, and off major river deltas, high nutrient supplies stimulate high rates of organic productivity by diatoms and dinoflagellates, often to the exclusion of coccolithophorids and planktic foraminifera, which reduces $CaCO_3$ production. At the same time, the rain of organic matter to the ocean floor supports abundant deep-sea life whose respiration adds significantly to the CO_2 in bottom waters. The result is substantial shoaling of the lysocline and CCD in these regions. The greater corrosiveness of AABW compared to NADW at approximately the same "age" is caused by upwelling-induced high organic productivity at the Antarctic divergence, which exports excess of organic matter into AABW.

Pelagic sediments in the Atlantic and Indian Oceans are predominantly calcareous oozes. In the Pacific Ocean, where the CCD is deeper, red clays dominate, especially in the North Pacific. Carbonate oozes delineate shallower regions in the south Pacific, including the East Pacific Rise and the complex topography to the southwest.

Siliceous Oozes

Biogenic siliceous oozes have two major and two minor contributors:

- Golden-brown algae known as diatoms (Bacillariophyceae) construct a type of shell called a frustule out of opalline silica.
- The radiolaria, a large group of marine protists distantly related to the foraminifera, also construct opalline silica skeletons.

Silicoflagellates are a minor group of marine algae that also construct opalline silica skeletons. Sponge spicules are also an important biogenic source of opalline silica in neretic waters, but are of minor importance in the deep sea.

Silica is undersaturated throughout most of the world's oceans. As a result, extraction of silica from seawater for production of silica shells or skeletons requires substantial energy. Furthermore, for siliceous sediments to be preserved they must be deposited in waters close to saturation with respect to silica and they must be buried quickly. Young seawaters that are highly undersaturated with respect to H_4SiO_4 are far more corrosive to SiO_2 than are old seawaters that have been dissolving and accumulating H_4SiO_4 over hundreds to thousands of years.

Seawaters around volcanic islands and island arcs tend to have higher concentrations of H_4SiO_4 in solution and therefore are more conducive to silica production in surface waters and silica preservation in sediments. Siliceous sediments are most common beneath upwelling zones and near high latitude island arcs, particularly in the Pacific and Antarctic. More than 75% of all oceanic silica accumulates on the sea floor between the Antarctic convergence and the Antarctic glacial marine sedimentation zone. Accumulation rates of siliceous oozes can reach 4-5 cm/1,000 years in these areas.

Conditions favoring deposition of silica or calcium carbonate are different. Silica solubility increases with decreasing pressure and increasing temperature. Silica is undersaturated in the oceans, but it is less undersaturated in deep water. Carbonate solubility increases with depth, and bottom waters become more undersaturated in calcium carbonate. The patterns of carbonate and silica deposits reflect different processes of formation and preservation, resulting in carbonates oozes that are poor in biogenic silica and vice versa.

The diatoms are extremely important primary producers that benefit physiologically from rich supplies of dissolved inorganic nutrients. Under such conditions, their growth rates far exceed other phytoplankton and they can rapidly produce both organic matter and siliceous sediments. They thrive in areas of intensive upwelling and near terrestrial sources of dissolved nutrients, including silica. Silicoflagellates show similar distributions. On the other hand, because both groups require substantial nutrient resources for growth, they are never abundant where nutrients are scarce, and so are insignificant primary and sediment producers in subtropical gyres. Diatom oozes, which contain more than 30% diatom frustules, are found beneath the Antarctic divergence, off the Aleutian island arc in the far North Pacific, and beneath areas of intensive meridional upwelling such as the eastern tropical Pacific. These oozes contain a significant percentage of radiolarian and silicoflagellate skeletons as well. Diatom-rich muds are common on continental shelves and margins where runoff from land contributes terrigenous muds as well as nutrients that stimulate diatom production.

Radiolaria, being protists, are slightly less dependent on the most nutrient-rich areas of the oceans. They are important contributors to siliceous oozes around the Antarctic, but radiolarian oozes (> 30% radiolarian skeletons) are primarily in the tropical Pacific beneath the equatorial upwelling zone and below the CCD. Above the CCD in this region, the sediments are calcareous with a significant siliceous component.

After burial, most siliceous oozes remain unconsolidated, but fractions dissolve and reprecipitate as chert beds or nodules. Chert is cryptocrystalline and microcrystalline quartz, which is very hard and impermeable. Chert beds are very difficult to drill, which has frustrated ocean drillers since the early days of the Deep Sea Drilling Project (DSDP). The abundance and widespread distribution of chert beds of Eocene age, discovered by the DSDP, indicate important changes in deep-sea chemistry over the past 50 million years.

Marine Sediments

Sediments are the earth's materials like gravel, sand, mud and lime that are transported and deposited by wind, water, ice and gravity. They get accumulated and cemented to form sedimentary rocks. Based on the mechanism of accumulation, sediments are classified into:

- Residual sediments.
- Mechanical sediments.
- Chemical sediments.
- Organic sediments.

Residual Sediments

Residual sediments are formed due to weathering. E.g. Laterite and bauxite.

- Mechanical sediments are formed due to deposition after transportation in water. E.g. Sandstone and conglomerates.

- Chemical sediments are formed due to precipitation of soluble constituents by water. E.g. Limestone.
- Organic sediments are formed due to the accumulation and preservation of organic matter.

Continuous sedimentation of organic and inorganic substances takes place in all aquatic ecosystems. Benthic fauna thrives using these resources. Bioturbation is the stirring of sediment by benthic fauna. This often destroys the structure of sediments. Recently laid sediments of lakes or seas may contain some detritus food and microorganisms. Benthic fauna, feed through burrowing of these recently laid sediments. The bedding and lamination structures already created in these sediments are disturbed. Marine sediments are formed by several processes. Marine sediments are brought in by the rivers, glaciers and wind along with shells and skeletons of organisms. Deep sea rocks that formed from these sediments are good source of information about the past sedimentary environments. Sediments are much useful in deciphering the source rock contribution and the locations of their origin. Sediments vary in their sizes, shapes, chemistry and density. Some are larger in size and some are very fine particles. Marine sediments are also classified based on their origin and particle size.

The Notable Categories of Marine Sediments

The notable categories of marine sediments are:

- Lithogenous Sediments or terrigenous sediments.
- Biogenous sediments.
- Hydrogenous sediments or authigenic sediments.
- Cosmogenous sediments.

Lithogenous Sediments

The term lithogenous means "derived from rocks". They originate from erosion of land, volcanic eruptions, and blown dusts. These mainly exist as continental masses, and as volcanic islands inside the oceans. These are important sources of lithogenous sediments. Weathering of minerals, on continental lithospheric rock bodies, provide the input for lithogenous sediments in the oceans. Quartz sands, clays, muds are the examples. Clays are the major products of mineral weathering reaction on land.

Terrigenous Sediments

Terrigenous sediments are those where the ultimate source is weathering and erosion of rocks on land. The materials composing these sediments are introduced to the ocean by water, wind or ice. Terrigenous sediments are more abundant close to the continents, specifically near river mouths and in the very deep areas of the oceans. The composition of terrigenous sediments thus depends on the climate and weathering at the source region. The major products of such processes are:

- Clays are derived from decomposition of feldspars.
- Chlorites are derived from decomposition of ferromagnesian minerals.
- Montmorillonite is a notable clay mineral.
- Illite is another clay mineral.
- Kaolinite is a product of Potash Felspar and mica.
- Zeolite is a hydrated aluminum silicate resulting the weathering of feldspars.

Wind plays a significant role in the transport of lithogenous sediments towards the oceans.

Biogenous Sediments

Biogenous sediments are derived from organisms. The materials include carbonate minerals created by organisms, such as corals, molluscs, and foraminifera. These are deposited at ocean floors with layers of calcite forming limestone. The other biogenic rocks include stromatolites, the flint nodules found in chalk and coal.

Bones, teeth and insoluble remains of organism are continuously deposited inside the oceans. The most common chemical compounds found as biogenous sediments are silica and calcium carbonate. Microscopic diatoms, animals like radiolarians and organisms like foraminifera are responsible for producing Calcium Carbonate component. Growing Algae and coral reef are the outcome of such biogenic sediment formation.

The Distribution of Biogeneous Sediments

The distribution of biogeneous sediments is controlled by three important processes:

- Production in surface waters: The growth of marine algae (which are the base of the oceanic food chain) is controlled principally by the availability of two critical nutrient elements -- N (nitrogen) and P (phosphorus).
- Dissolution in deep waters: Deep-ocean waters are undersaturated in calcium carbonate and opalline silica. Therefore, biogenic particles tend to dissolve as they settle through the water column and as they sit on sea floor.
- Dilution on the sea floor: Calcareous and siliceous components can be diluted to < 30 % of the total sediment (and therfore not qualify as a "biogenous ooze") in regions where the input of terrigenous sediment is very high.

Hydrogenous Sediments

Hydrogenous sediments are mainly derived from water. They are formed by the chemical reaction and precipitation that occur within the seawater. Sea water contains dissolved ions (about 3.5 % by weight). Salts precipitate from "super-saturated" sea water in restricted basins and are deposited on the sea floor.

The notable minerals formed under this category are Manganese deposits, Phosphorites and Glauconites.

"Manganese nodules" are oxides of the metals manganese (Mn) and iron (Fe). They also contain minor amounts of copper (Cu), cobalt, (Co), and nickel (Ni). Manganese nodules occur as nodules and crusts, primarily in deepest of oceans and near mid-ocean ridges. The metals may be delivered as dissolved species or as sediments from land.

The rate of accumulation of these hydrogenous sediments is very slow. One of the prominent marine sediments having great economic value is the occurrence of Manganese nodules. These are polymetallic nodules. They are found in deep ocean floors. The major components of manganese nodules are Mno and Fe_2O_3. In addition, they contain copper, cobalt and nickel. The genetic cause of these nodules may be related to the volcanic products, runoff from continents and other mineral-weathering reactions happening on rocks.

Bacteria are an active player in the formation of manganese nodules. The other marine sediments include a good amount of Phosphorites, Glauconites and Carbonates.

Cosmogenous Sediments

Cosmogenous sediment is sediment that is derived from outside the earth. This is usually in the context of minor stardust or broken down asteroid particles. One can usually identify an accumulation of the cosmogenous sediment in areas where there is minor terrigenous sediment and by testing the isotopes. These sediments accumulate in the oceans (and land) from "micro-meteorites" whose composition can be either silicate (like the mantle), metal (like the core), or a mixture of the two. Although there is a continuous "rain" of these particles on Earth, cosmogenous sediments make up a very minor component of ocean sediments.

Composition of Marine Sediments

Marine sediments are of two major categories as:

- Neritic sediments.
- Oceanic sediments.

Neritic sediments are those that are deposited near the continents and oceanic sediments are the characteristic of deep-ocean basins.

The major outputs of these are:

- Turbidites.
- Glacial deposits.
- Carbonate deposits.

The sediments of deep ocean basins include:

- Abyssal clay.
- Calcareous and Siliceous oozes.

Neritic sediments refer to mainly lithogenous sediments. Oceanic sediments refer to organic and hydrogenic categories of sediments. Marine sediments are called as Pelagic sediments when they accumulate in the abyssal plain of the deep ocean, far away from terrestrial sources. Pelagic sediments are primarily limited to the continental shelf. Pelagic sediments that are mixed with terrigenous sediments are known as hemipelagic.

There are three main types of pelagic sediments:

- Siliceous oozes.
- Calcareous oozes.
- Red clays.

Ooze does not refer to sediment's consistency, but to its origin. Oozes are primarily biogenic that is, composed of planktonic debris while red clays are non-biogenic, with little organic material.

Any sediment that is composed of more than 30% of microscopic skeletal debris is classified as an ooze. The type of sediment that accumulates at a given locale is determined by the location's distance from land, water depth, and overall fertility. Oozes are not present on the continental margin.

1. The Calcareous ooze is composed primarily of the shells--also known as tests--of foraminifera, coccolithophores, and pteropods. This is the most common pelagic sediment by area, covering 48% of the world ocean's floor. This type of ooze is limited to depths above the Carbonate Compensation Depth at time of burial. It accumulates more rapidly than any other pelagic sediment type, with a rate that varies from 0.3 - 5 cm / 1000 yr.

2. The Siliceous ooze is composed of the debris of plankton with silica shells, such as diatoms and radiolaria. This ooze is limited to areas with high biological productivity, such as the polar oceans, and upwelling zones near the equator. The least common type

of sediment, it covers only 15% of the ocean floor. It accumulates at a slower rate than calcareous ooze: 0.2-1 cm / 1000 yr.

3. The Red clays are also known as pelagic clays. They accumulate in the deepest and most remote areas of the ocean. They contain less than 30% biogenic material, their composition is a varied. There will be a mix of very fine quartz and clay minerals. The authigenic deposits precipitated directly from seawater, and also from micrometeorites.

The red clays take the color of oxidized iron minerals. Sometimes, it is brownish in color. Its ultimate origin is uncertain. Red clay seems to be mostly derived from distant rivers, and windblown dust. These deposits cover about 38% of the ocean floor. They accumulate more slowly than any other sediment type, at only 0.1-0.5 cm / 1000 yr.

There are several kinds of oozes recognized based on their content and composition. The notable ones are:

- Diatom ooze.
- Radiolarian ooze.
- Foraminifera ooze.
- Pteropd ooze.
- Globigerina ooze occur in deep oceans.

Destruction of skeletal remain occurs due to solution. Dilution is a major process of oceanic waters. The origins of Marine sediments are easily indentified from their mineral composition. Occurrence of rock-farming minerals like quartz and feldspar are indicative of their origin from rocks of the land. Occurrence of silica (SiO_2) and calcite ($CaCo_3$) may be attributed to the compound derived from organisms. Numerous minerals also get precipitated from ocean waters. Phosphorites are found at depths above 1000m and Glauconites are found at a depth of 2500m.

Aragonites and Calcites are the minerals of Carbonate group. Precipitation of these minerals happens in the form of Oolites. Due to the occurrence of much carbonate masses, oceans are considered to be the reservoirs of carbonates and a store house of carbon-di-oxide.

The cosmogenous sediments are of cosmic in origin. These are the microscopic spherules rich in iron and rocky chondrules. Meteoric impacts are the major reasons for the occurrence of such sediments in oceans. Metallic element like iridium is found in marine sediments, which is indicative of cosmogenic origin.

Distribution of Oceanic Sediments

There are two aspects related to marine sedimentation. One is the settling rate and the other one is the sedimentation rate. Water can carry the sediments as suspended load and bed load. These sediments are settled when there is a change in the velocity of flow or due to any obstruction for movement.

The settling velocity is the minimum velocity a water flow must have in order to transport the sediments rather than depositing them. This is given by the Stoke's Law:

SV= 2(Dp-Df) gr^2/ 9 m

Where,

- SV is the settling velocity,
- D is density,
- p indicates particle,
- f indicates the fluid,
- g is the acceleration due to gravity,
- r is the radius of the particle,
- m is the dynamic viscosity of the fluid.

If the flow velocity is greater than the settling velocity, sediment will be transported. Otherwise, they are settled. The rate of sedimentation depends on various factors like daily input, seasonal inputs and annual inputs from various sources.

Oceanic sediments cover about 75% of the ocean floors. In this, the calcareous oozes alone cover about 48% of the deep sea floors. Abyssal clay covers about 38% and siliceous oozes cover about 14% of the total area. Calcareous oozes are the dominant oceanic sediments of the Indian and Atlantic oceans. Abyssal clays are the principal marine deposits of the Pacific Ocean.

One of the major marine sediments confusing the oceanographic studies is due to the occurrence of Fecal Pellets. These are derived from the surface directly above the seas and oceans. Fecal Pellets produced from tiny animals living in water are also the major contributors.

The following are the statistical observations:

- The rate of accumulation of lithogenous particles in 1.27 cu.km/year. Biogenous particles 0.38 cu.km/year. Both of them put together contribute to 1.65 cu.km/year.
- The rate of volcanic sediment accumulation is 0.05 cu.km/year.
- Seamounts add 0.2 cu.km/year of mass to the ocean crust and volcanic islands add 1.1 cu.km/year.

All of these put together contribute 3 cu.km/year of sediments into the ocean floors.

Shales, Argillite and Siltstone

Argillaceous/mud rocks are a class of fine grained siliciclastic sedimentary rocks composed of at least 50% silt- and clay-sized particles. These relatively fine-grained particles are commonly transported as suspended particles by turbulent flow in water or air and deposited as the flow calms and the particles settle out of suspension. The varying types of mud rocks include: siltstone, claystone, mudstone, slate and shale. Most of the particles are less than 0.0625 mm (1/16th mm or 0.0025 inches) and are too small to study readily in the field. At first sight the rock types look quite similar; however, there are important differences in composition and nomenclature. There has been a great deal of disagreement involving the classification of mud rocks. There are a few important hurdles to classification, notably:

- Mud rocks are the least understood and one of the most understudied sedimentary rocks to date.
- It is difficult to study mud rock constituents due to their diminutive size and susceptibility to weathering on outcrops.
- There is more than one classification scheme accepted by earth scientists.

Shales are most abundant sedimentary rock forming about one half of the total sedimentary rocks. 'Clay' is natural plastic earth composed of hydrous aluminium silicates and sediment grains less than 0.002mm (1/256mm) in diameter.

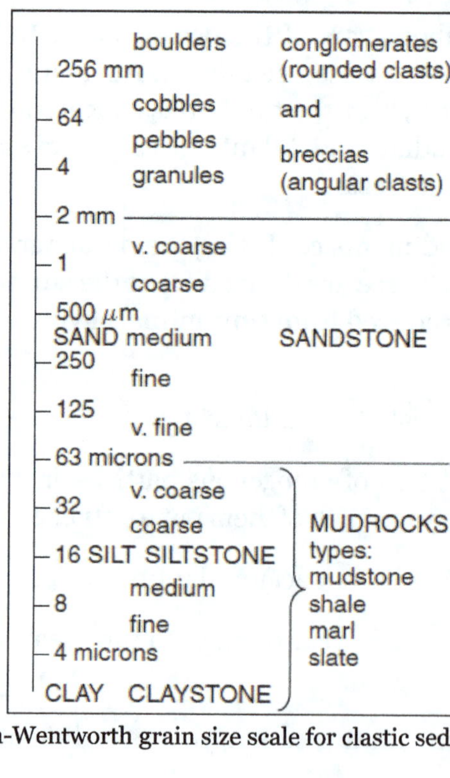

Udden-Wentworth grain size scale for clastic sediments.

Claystone is indurate clay. If it possesses bedding fissility then it is called shale. Shale is a laminated, fissile rock. Claystone which are neither fissile nor laminated but are blocky and massive, term mudstone is commonly used.

Shale

Claystone

Siltstone is indurate silt ranging in particle size from 1/16 to 1/256mm in diameter or sediments in which 50 percent or more of the particles fall in this category. Unlike shales, siltstones are generally held together by chemical cements and may show small scale cross bedding, convolute bedding etc.

Siltstone

Term 'argillite' is used in several ways. A rock derived from siltstone or shale that has undergone a somewhat higher degree of induration than normally present in siltstone and shale. That is, argillite is intermediate in character between shale and slate. Some researchers the term 'argillite' for clay or shale hardened by recrystallisation and called it slate if it possesses secondary cleavage. Shales are typically composed of variable amounts of clay minerals and quartz grains and the typical colour is gray. Addition of variable amounts of minor constituents alters the colon of the rock. Black shale results from the presence of greater than one percent carbonaceous material and indicates a reducing environment. It is good source rock for hydrocarbons and can contain up to twenty percent organic carbon. Generally, black shale receives its influx of carbon from algae, which decays and forms ooze known as sapropel. When this ooze is cooked at desired pressure, three to six kilometres depth, and temperature, 90-120 degrees Celsius, it will form kerogen. Red, brown and green colours are indicative of ferric oxide (hematite – reds), iron hydroxide (goethite – browns and limonite – yellow), or micaceous minerals (chlorite, biotite and illite – greens).

The process in the rock cycle which forms shale is called compaction. The fine particles that compose shale can remain suspended in water long after the larger particles of sand have deposited. Shales are typically deposited in very slow moving water and are often found in lakes and lagoon deposits, in river deltas, on floodplains and offshore from beach sands. They can also be deposited on the continental shelf, in relatively deep, quiet water. Fossils, animal tracks/burrows and even raindrop impact craters are sometimes preserved on shale bedding surfaces. Shales may also contain concretions consisting of pyrite, apatite, or various carbonate minerals.

A claystone is lithified and non-fissile mud rock. A rock to be considered as claystone, must consists of up to fifty percent clay, which measures less than 1/256 of a millimetre in particle size. Clay minerals are integral to mud rocks and represent the first or second most abundant constituent by volume. There are 35 recognized clay mineral species on Earth. They make muds cohesive and plastic, or able to flow. Clay is by far the smallest particles recognized in mud rocks. Most materials in nature are clay minerals but quartz, feldspar, iron oxides, and carbonates can weather to sizes of a typical clay mineral. For a size comparison, a clay-sized particle is 1/1000 of a sand grain. This means a clay particle will travel 1000 times further at constant water velocity, thus requiring quieter conditions for settlement.

The formation of clay is well understood, and can come from soil, volcanic ash and glaciations. Ancient mud rocks are another source, because they weather and disintegrate easily. Feldspar, amphiboles, pyroxenes and volcanic glass are the principle donors of clay minerals. One of the highest proportions of silt found on Earth is in the Himalayas, where phyllites are exposed to rainfall of up to five to ten meters a year. Quartz and feldspar are the biggest contributors to the silt realm and silt tends to be non-cohesive, nonplastic, but can liquefy easily.

Heavy rainfall provides the kinetic motion necessary for mud, clay, and silt transport. Southeast Asia, receives high amounts of rain from monsoons, which then washes sediment from the Himalayas and surrounding areas to the Indian Ocean. Warm, wet climates are best for weathering rocks, and there is more mud on ocean shelves off tropical coasts than on temperate or polar shelves. Rivers, waves, and long shore currents segregate mud, silt and clay from sand and gravel due to fall velocity. Longer rivers, with low gradients and large watersheds, have the best carrying capacity for mud.

Mud rocks, especially black shale, are the source and containers of precious petroleum sources throughout the world. Since mud rocks and organic material require quiet water conditions for deposition, mud rocks are the most likely resource for petroleum. Mud rocks have low porosity, they are impermeable and often, if the mud rock is not black shale, it remains useful as a seal to petroleum and natural gas reservoirs.

Argillaceous rocks (also called lutaceous) are also economically important and many clays and shales are used as raw materials for manufacturing of bricks, building tiles, roofing and ceramic wares. Clays are also used to form Portland cement. Slate, formed by metamorphism of shale, is used as blackboards, roofing tiles and electrical panels.

Texture and Structure

Grain Size

Because of fine grained character particle size of clays is determined by methods based on differential settling velocities. Average shale contains about two parts silt and one part clay. Texturally, clay is defined as all material finer than 4 microns; silt ranges in size from 4 to 63 microns; sand ranges from 63 microns to 2 mm.

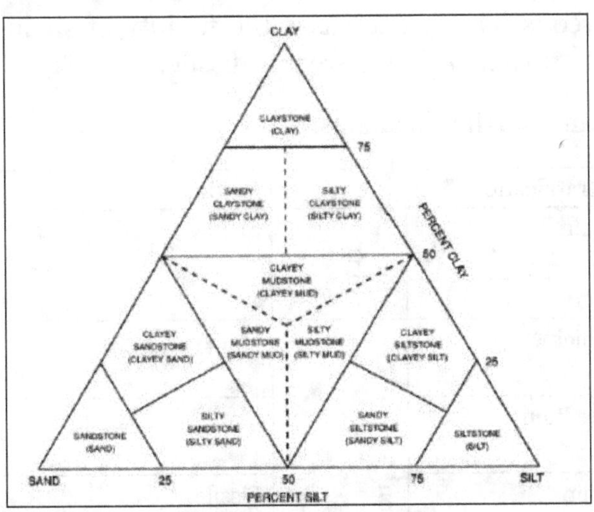

Textural classification of fine-grained rocks and sediments.

The sizes of fine, unconsolidated particles can be determined by using a number of instrumental techniques, which are based mainly on measuring the settling velocity of particles in water. These methods cannot be used to measure the grain sizes of particles in mudstones and shales, which generally cannot be disaggregated to yield individual particles. The sizes of grains in consolidated rocks can be measured by using an electron microscope.

Particle Shape

The shapes of the small particles that make up mudstones, unlike the shapes of sand-size and larger particles are little modified by sediment erosion and transport. Very small quartz particles (< ~ 0.1mm) do not become rounded very effectively by any type of eolian or stream transport. Therefore, the shapes of finesilt- and clay-size particles in mudstones reflect mainly the original shapes of the detrital particles, largely unmodified by transport abrasion, or they reflect the shapes of minerals generated during diagenesis. Thus, most particles in mudstones are very angular. Many particles, especially clay minerals and fine micas have very low sphericity. Electron microscopy reveals that most clay minerals have platy, flaky or acicular shapes.

Fissility

Fissility is tendency of a rock to split or separate along relatively smooth surfaces parallel to the bedding. Shales contain high concentrations of clay minerals that have platy or flaky shapes. These rocks may exhibit microfabrics resulting from the preferred orientation of flaky clay minerals.

A fissile fine-grained rock is one that tends to split relatively easily into thin, approximately parallel layers that range in thickness from about 0.5mm to 1.0 mm. Shales that split into thinner layers are called papery shales, and those that split into thicker units are platy, flaggy, or slabby shales depending upon thickness of the parting. An increase in siliceous or calcareous matter decreases the fissility of shale while carbonaceous matter increases it. Bioturbated shales are non-fissile.

Table: Stratification and parting in shales.

<table>
<tr><th>Thickness</th><th colspan="2">Stratification</th><th>Parting</th><th>Composition</th></tr>
<tr><td>30 cm</td><td>Thin</td><td rowspan="2">Bed-ding</td><td rowspan="2">Slabby</td><td rowspan="6">----Clay and organic content---->
----Sand, silt, and carbonate con-tent---></td></tr>
<tr><td>3 cm</td><td>Very Thin</td></tr>
<tr><td>10 mm</td><td>Thick</td><td rowspan="4">Lamination</td><td>Flaggy</td></tr>
<tr><td>5 mm</td><td>Medium</td><td>Platy</td></tr>
<tr><td>1 mm</td><td>Thin</td><td>Fissile</td></tr>
<tr><td>0.5 mm</td><td>Very Thin</td><td>Papery</td></tr>
</table>

Laminations

Laminations of shales range from 0.05 to 1.0 mm in thickness in the form of alternations of coarse and fine particles like silt and clay, light and dark layers and calcium carbonate and silt. This happens mainly due to differential settling velocity of different constituents or different rate of supply of these materials at the site of deposition. Uniform sedimentation over a long period of time may produce structureless sediments.

Other Structures

Shales and siltstones also contain concretionary bodies like calcareous concretions, chert nodules, iron concretions etc.

Microfabric

Three kinds of depositional and diagenetic processes and mechanisms may operate to produce clay and shale microfabrics: physicochemical, bioorganic and burial-diagenesis. In turn, physicochemical processes take place by three mechanisms: electrochemical (forces that hold particles together internally and that bind particles), thermochemical (forces arising from temperature and temperature differences) and interface dynamics (differential motion of settling particles under the influence of gravity, differential flow of water masses of differing density, impact of particles on sediment interface, flow at the interface, and micro roughness of the interface). Bioorganic processes represent the effects of living organisms on sediment properties and are brought about by biomechanical (bioturbation), biophysical (aggregation or agglutination of particles by organic processes) and biochemical (chemical production and destruction of chemical entities, e.g. production of gases by organisms) mechanisms. Burial-diagenesis processes that can affect shale microfacies take place by mass gravity mechanisms (mechanical rearrangement of particles owing to overburden stresses) and cementation and other diagenetic phenomena.

Composition

Composition of clays and shales consist of material produced by abrasion (mainly silt) and weathering (residual clays) along with chemical and biochemical additions. Chemical additions are either the products of precipitation from solution or deposition along with associated clays.

Clay minerals, fine-size micas, quartz and feldspars are the most abundant minerals in mudstones and shales. A variety of other minerals may occur in these rocks in minor quantities, including zeolites, iron oxides, heavy minerals, carbonates, sulfates and sulfides, as well as, fine-size organic matter. Because of the difficulties involved in petrographic analysis of fine sediments, most investigators have concentrated on the clay mineralogy of shales, which can be determined fairly easily by X-ray diffraction methods. Quantitative to semi-quantitative determination of fine quartz, feldspars, and other minerals in shales can be made also by X-ray diffraction methods with a scanning electron microscope equipped with an energy-dispersive X-ray unit or with an electron probe microanalyzer.

Variations in mineral abundances may be due to several factors like mineral composition is known to vary markedly with grain size. Quartz tends to be more abundant in coarser-grained mudstones and shales, whereas clay minerals are more abundant in finer-grained mudstones and shales. Principal constituents in shales and mudstones are tabulated in table.

Table: Principal constituents in shales and mudstones.

Constituents	Remarks
Silicate minerals	
Quartz	Makes up 20 to 30 percent of the average shale; probably mostly detrital; chalcedony and biogenic may also be present.
Feldspars	Commonly less abundant than quartz; plagioclase generally more abundant than alkali feldspars.
Zeolites	Commonly present as an alteration product of volcanic glass; phillipsite and clinoptilolite are common zeolites in modern marine sediments.
Clay minerals	
Kaolinite (7 Å)	Forms under strongly leaching conditions: abundant rainfall, good drainage, acid waters; in marine basin tends to be concentrated near shore.
Smectite–illite–muscovite (10 Å and greater)	Smectite is a hydrated, expandable clay; common in soils and as an alteration product of volcanic glass; alters to illite during burial; illite is the most abundant clay mineral in shales; derived mainly from pre-existing shales; alters to muscovite during diagenesis; muscovite may also be detrital.
Chlorite, corrensite, and vermiculite	Chlorite forms particularly during burial diagenesis; second in abundance only to illite in Paleozoic and older shales; during burial, vermiculite may convert to corrensite and finally to chlorite.
Sepiolite and attapulgite	Magnesium-rich clays that form under special conditions where waters pore are rich in Mg, e.g. saline lakes.
Oxides and hydroxides	
Iron oxides	Hematite most common in shales, but goethite or limonite may be more common in modern muds; commonly present as coatings on clay minerals; may be converted to pyrite or siderite in reducing environments.
Gibbsite	Consists of $Al(OH)_3$; may be associated with kaolinite in marine shales derived from the weathering of tropical landmasses.

Carbonates	
Calcite	More common in marine than non-marine shales.
Dolomite	An important cementing agent in some shales.
Siderite and ankerite	Occur in shales most commonly as concretions.
Sulfur minerals	
Sulfates: gypsum, anhydrite, and barite	Occur as concretions in shales and may indicate the presence of hypersaline conditions during or after deposition.
Sulfides	Mainly the iron sulfides pyrite and marcasite; these sulfides are most abundant in marine shales and indicate reducing conditions either at the time of deposition or during diagensis.
Other constituents	
Apatite	Occurs particularly as nodules in marine shales that accumulated slowly in areas of high organic productivity.
Volcanic glass	Common in modern continental and marine muds in areas of volcanic activity; converts to zeolites and smectities during burial diagenesis.
Heavy minerals	Occur in shales, but little is known about patterns of occurrence and relative abundance.
Organic substances	
Discrete and structured organic particles	Mostly palynomorphs or small coaly fragments (vitrinite).
Kerogen	Occurs in all shales except red ones.

When silicates of primary crystalline rocks are decomposed by weathering, clay minerals are formed as resultant product. Clay minerals are hydrated silicates of aluminium with some replacements by iron and magnesium. Clay minerals are fine grained, usually less than 5 micron in size and even in millimicron.

Clay minerals belong to the group of silicate minerals known as phyllosilicates. They are characterized particularly by SiO4^{4-} ionic groups in combination with metallic cations. The SiO4^{4-} groups consist of a silicon atom surrounded by four oxygen atoms in a tetrahedral configuration. Therefore, they are called silica tetrahedra or silicon –oxygen tetrahedra. Silica tetrahedra can be linked together to form indefinitely extending tetrahedral sheets.

Common clay minerals of shale are phyllosilicates i.e. they have sheet structure consisting of two types of layers. One is silica tetrahedral layer and another is aluminium hydroxide. Clay minerals belong to two groups.

- Kaolinite group: Mineral is characterised by two layer (1:1 layer) lattice consisting of one octahedral or gibbsite layer linked to one silica tetrahedral layer. The lattice does not expand with varying water content and no replacement by iron or magnesium in the gibbsite layer.

- Another group is three layer lattice (2:1) lattice: In this group alumina octahedral layer is sandwiched between two silica tetrahedral layer.

The clay minerals are represented largely by kaolinite, montmorillonite and illite. Clay minerals of Late Tertiary mudstones are expandable smectites whereas in older rocks especially in mid to early Paleozoic shales illites predominate. The transformation of smectite to illite produces silica, sodium, calcium, magnesium, iron and water. These released elements form authigenic quartz, chert, calcite, dolomite, ankerite, hematite and albite, all trace to minor (except quartz) minerals found in shales and other mudrocks.

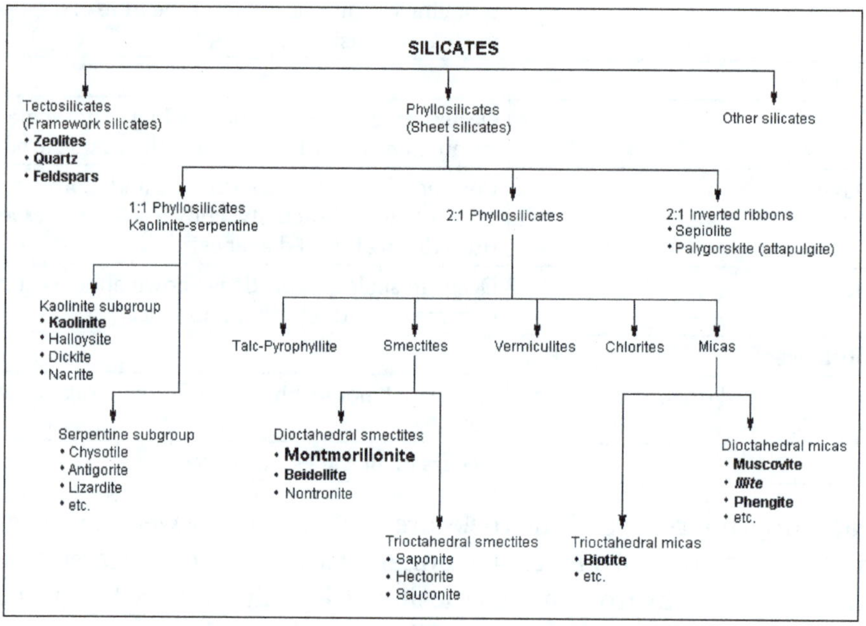

Different clay minerals derived from silicates.

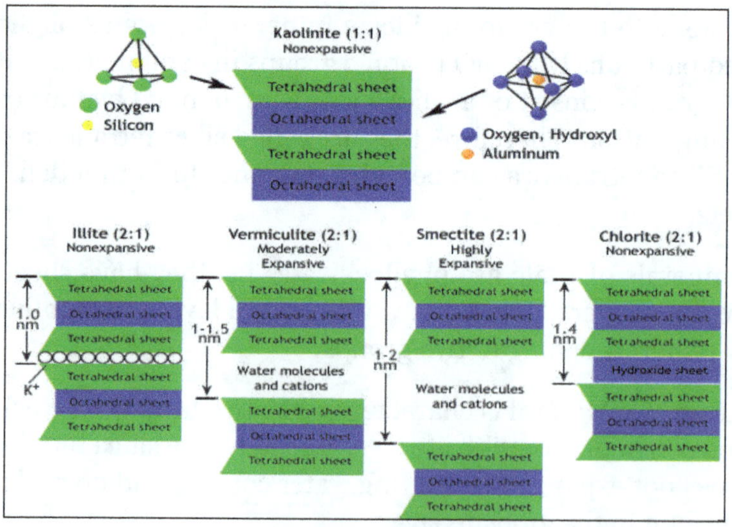

Structure of clays.

Shales and mud rocks contain roughly 95 percent of the organic matter in all sedimentary rocks. However, this amounts to less than one percent by mass in average shale. Black shales which form in anoxic conditions contain reduced free carbon along with ferrous iron (Fe^{2+}) and sulfur (S^{2-}). Pyrite and amorphous iron sulfide along with carbon produce the black coloration and purple.

'Black shales' are dark, as a result of being especially rich in unoxidized carbon. Common in some Paleozoic and Mesozoic strata, black shales were deposited in anoxic, reducing environments, such as in stagnant water columns. Some black shales contain abundant heavy metals such as molybdenum, uranium, vanadium, and zinc.

Shales that are subject to heat and pressure of metamorphism alter into a hard, fissile, metamorphic rock known as slate. With continued increase in metamorphic grade the sequence is phyllite, then schist and finally to gneiss.

Diagenesis

Shales are subject to many post- depositional changes, physical as well as chemical. It involves compaction (pore space reduction and change of clay minerals orientation) and mineral transformation by surrounding medium or recrystallisation.

Porosity of freshly deposited mud is quite high even exceeding 50% but porosity of shale is quite low. This decrease in porosity and conversion of mud to shale is largely due to compaction by superincumbent load. Chemical and mineralogical changes start soon after deposition of sediments and continue along with aging of sediments including recrystallisation with increasing temperature and pressure. Clay minerals generated during weathering are very susceptible to changes in marine environment. Clay minerals are environment sensitive and can be used to interpret paleoenvironmental conditions. Any of the clay mineral can occur in any of the major depositional environment. Kaolinitic clays after reaching the marine environment are altered to iillite and chlorite and montmorillonite is converted to chlorite. Clay mineralogy of sediment may be altered on burial (with rise in temperature and pressure). Montmorillonite and kaolinite tend to disappear with increasing depth and change to illite and chlorite.

Distribution and Significance

Occurrence in Time and Space

Mud stones and shales make up about 50 percent of all sedimentary rocks. Most of these mudstones and shales are of marine origin. They occur in stratigraphic successions of all ages and on all continents of the world. Shales have been the dominant sedimentary rock deposited throughout most of geologic time and overall abundance in the geologic record reflects the high average abundance of fine-grained material generated by the processes of weathering and erosion (which are related to elevated landmasses and high rainfall). The wide distribution of mudstones and shales in space and time suggests that the environments that favor deposition of fine sediment are common and that they have recurred throughout time. Mudstones and shales are not especially characteristic of any particular geologic time. They are abundant in rocks of all ages but geographic distribution varies as a function of age. Thus, their relative abundances reflect changing patterns of land and seas through time. The relative abundances of mudstones and shales in stratigraphic successions of a particular age are related to depositional setting. The thickest shale and mudstone sections tend to occur in sedimentary successions that were deposited in mobile, marine basins. Nonetheless, moderate thicknesses of non-marine mudstones and shales can occur in stratigraphic successions deposited in interior basins, either in lacustrine or alluvial basins.

Ancient muddy sediments were deposited particularly in meander and anastomosing stream systems on continents; in inter distributary portions of deltas, in some quiet-water transitional-marine coastal environments such as lagoons and estuaries, and in many low-energy marine environments on the shelf, slope, and deep ocean floor. Fine grained sediment may be deposited on land by wind to form loess or other eolian deposits; however, most fine, siliciclastic sediment is deposited in water. Deposition takes place mainly in quiet water, below wave base, at water depths ranging from tens of meters to thousands of meters. Such low-energy environments favoring deposition of argillaceous sediments can be present in a variety of depositional settings on continents (i.e. lakes, alluvial piedmonts or plains, river floodplains), in transitional marine environments (deltas, estuaries, lagoons, tidal flats), on the shallowmarine shelf and the continental slope, and in deeper-marine basins. In continental settings, fine sediment is transported in suspension by wind and by stream flow. It may also be transported, together with sand and gravel, by glacial ice and by mass-transport processes (mud flows, debris flows, turbidity currents in lakes). Fine sediment that is delivered to the ocean by any of these processes can be re-transported further seaward by several mechanisms that may include storm-generated waves and currents, suspension transport in nearsurface plumes, near-bottom nepheloid transport, turbidity currents, and bottom currents. Overall, turbidity currents are probably the most important mechanism for moving large quantities of fine sediment to distal parts of deep basins. Wind may also transport fine sediment from land to distant regions of the ocean and may be the principal mechanism for transport of pelagic sediment.

Within the depositional environment, deposition of fine sediment takes place by slow gravity settling of dispersed single particles or by more-rapid settling of clay flocs. Organisms may aid in deposition of fine sediment by pelletizing muds and by creating baffles or other features (e.g. algal mats) that trap and hold fine sediment.

Mudstones and shales deposited in a given depositional environment acquire characteristic properties (geometry, lithology, facies associations, textures, structures, fossil content) that may allow them to be distinguished from mudstones and shales of other environments.

Significance of Mudstone and Shale

The presence of thick units of mudstones or shales in ancient stratigraphic sequences implies derivation of fine-grained sediment from large-volume land masses that were weathered and eroded under generally humid, high-rainfall conditions. Furthermore, the presence of mudstones and shales in a stratigraphic section suggests deposition in a quiet-water paleoenvironmental conditions.

Most thick units were deposited in marine basins and tend to be laterally extensive. Individual marine mudstone and shale beds may be relatively thin, but they can commonly be traced laterally for considerable distances. By contrast, associated sandstone, conglomerate or limestone facies are generally more restricted in their lateral extent. Therefore, many mudstone and shale beds make excellent stratigraphic markers. In comparison to marine mudstones and shales, shale and mudstone sequences deposited in nonmarine settings tend to be thinner overall and laterally less extensive. Mudstones and shales may display lithologic characteristics, sedimentary structures, fossil content, organic content, geochemical characteristics and facies associations that have environmental significance.

Mudstones and shales, or facies closely associated with these rocks, may contain directional sedimentary structures that are useful in paleocurrent analysis. Clay minerals and feldspars in fine-grained rocks may have some importance as provenance indicators, although these rocks are generally less useful than sandstones in provenance studies. Moreover, mudstones and shales have considerable economic significance as source beds for petroleum and natural gas. Careful geochemical evaluation of the amount and types of kerogen in fine-grained rocks is now routine procedure in petroleum exploration excursions. Thus, in summary, mudstones and shales have paleoclimatic, paleoenvironmental, provenance and economic significance.

Sandstone

Sand/Sandstone is among the most common sedimentary rock and found in the entire geological record. This among the most widely studied rocks and it has proved its importance in various forms for the entire civilisation. Though there are diverse views about total abundance of sandstones in stratigraphic record but about 20% of all sedimentary rocks are generally believed.

Sand is loose, non-cohesive granular material and term is defined by its size. Grain size ranging from 0.0625 (1/16) mm to 2mm in diameter. Sandstone is lithified rock as a result of diagenesis. It is found in the entire depositional environment ranging from sand dunes, lakes, river channels, flood plains, backwater swamps, beaches, shoreline, lagoons, and tidal flats to deep water.

Sandstone section.

Composition

Sandstones are mixtures of mineral grains and rock fragments coming from naturally disaggregated products of erosion of all types of rocks. In this way, total types of minerals found in the sandstones are very large and principally controlled by the provenance. Hence, sandstones are important to understand the provenance including tectonics and climate.

Original minerals from source area may be lost or modified during the process of weathering, transportation or diagenesis. In the process of weathering minerals are altered by chemical decomposition like feldspar may alter to kaolinite. But some minerals like quartz which is very hard and resistant survive the chemical decomposition. During transportation, attrition of minerals causes decrease in size. Due to differential abrasion softer minerals decrease more in size and become more rounded than their harder counterparts. Consequently, sand sized grains are reduced to silt size or clay size. Mineralogical composition of sandstones is also changed to large extent by dissolution, precipitation and alteration during the diagenesis. During diagenesis, unstable minerals are lost and new minerals like carbonate may precipitate from the solution. Size of mineral crystals in source rock determines the presence of monomineralic or rock fragments in the resultant sandstone. Source rock with grain size less than that of sand (<1/16mm) will contribute as rock fragments in sandstone.

By definition, terrigenous sedimentary rocks composed mainly sand size particles are called sandstone. Sandstones may be deposited in variety of environments ranging from marine to non-marine and mixed. Main constituents of sandstones are sand sized quartz, feldspars, micas, heavy minerals and rock fragments forming the bulk of framework grains. The voids among the framework grains are filled by finer material (matrix, <0.03mm) or by chemically precipitated cement. Cement shows crystalline texture while matrix is composed of finer detrital particles.

Mineralogy of framework grains in sandstone can be analysed with standard petrographic microscope (study under thin sections) or by backscattered electron microscopy. Bulk chemical composition can be determined by techniques like X-ray fluorescence and inductively coupled argon plasma emission spectrometry (ICP). The chemistry of individual grains is determined by use of an electron probe analyser or an energy dispersive X-ray detector (EDX) attached to a scanning electron microscope (SEM).

Mineralogical Composition of Sandstones

Detrital Minerals

1. Quartz: Crystalline polymorph of SiO_2 namely quartz is thermodynamically stable under sedimentary conditions and it is one of the common minerals in sandstones. Other polymorphs like tridymite and cristobalite are not common. Quartz makes up 50-60% of framework fractions in sandstones. It is comparatively easy to identify in hand specimens, as well as, in thin sections. Because of its hardness and chemical stability it can survive multiple recycling. Quartz may occur as single (monocrystalline) grain or as composite (polycrystalline) grains. Quartz is derived from plutonic igneous rock, as well as, from metamorphic and older sedimentary rocks. The varieties of quartz most useful for source evaluation are polycrystalline quartz grains, chert grains and grains made up of quartz and rounded secondary overgrowth that can be recognised as second cycle grains. Quartz is also abundant as chemical cement and found commonly in sandstones. This type of quartz is relatively free of inclusions and trace elements and indicates low temperature origin. Chert and opaline silica are also found as pore filling cements or as rims around detrital quartz grains.

Table: Common minerals of sedimentary rocks.

Type of minerals	More than 10% of rock	Less than 10% of rock	Less than 1% of rock (Accessory minerals)
Detrital minerals	Quartz Clay minerals Mica (Fine)	Detrital chert Coarser mica Feldspar	Zircon, Tourmaline Epidote, Garnet Hornblende, Pyrite
Chemical and authigenic minerals	Calcite, Dolomite	Chert, Gypsum	Anatase, Feldspar, Mica

2. Feldspars: All varieties of feldspars are considered as detrital minerals and feldspar grains make up 10-20% of framework grains of average sandstones. After quartz, feldspars are second most abundant minerals. Several varieties of feldspars are distinguished on the basis of chemical composition and optical properties. Feldspars have been studied in detail for mineralogical, crystallographic and petrologic significance.

Feldspars are chemically less stable than quartz and more susceptible to chemical alteration during weathering and diagenesis owing to their softer nature, prone to mechanical shattering and cleavage. It is difficult on the part of feldspars to survive multiple cycles and hence if feldspar content of the rock is around 25% then it can be deduced that particular rock is derived directly from the crystalline source rocks.

Alkali feldspars constitute a group of minerals ranging in chemical composition from $KAlSi_3O_8$ through (K, Na) $AlSi_3O_8$ to $NaAlSi_3O_8$. Because of dominance of potassium rich feldspars, alkali feldspars are generally also called potassium feldspars (K- feldspar). Important member of this group are orthoclase, microcline and sanidine. Plagioclase feldspars form a complex solid solution series ranging in composition from $NaAlSi_3O_8$ (albite) to $CaAlSi_3O_8$ (anorthite). General formula for this series is (Na, Ca) (Al, Si) Si_2O_8. Plagioclase feldspars can be distinguished from alkali on the basis of optical properties. In general, alkali feldspars are more common in sedimentary rocks and plagioclase is more common in sandstones derived from volcanic rocks.

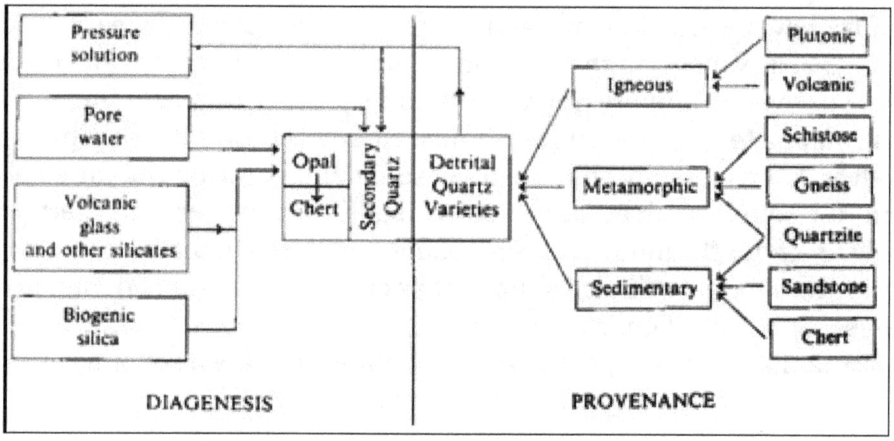

Figure: Origin of silicate minerals in sandstones.

Though feldspars are commonly found in metamorphic and igneous rocks similar to quartz and origin of feldspar can be deduced only after precise composition is defined. Composition can be determined by optical study, X-ray diffraction or staining techniques. In general, sanidine is associated with high temperature contact metamorphic rocks or volcanic rocks. Contrary to this microcline is widely present in metamorphic and plutonic igneous rocks but not in volcanic rocks. Plagioclase is not diagnostic of specific rock types and found in both igneous and metamorphic rocks. Abundance of feldspar is important to know source rock composition, chemical weathering in source area, abrasion and solution during transportation and solution during diagenesis. Authigenic feldspars in sandstones have been studied by various workers. The chemical conditions for precipitation of feldspars are determined by relative amounts of their components in solution and pH.

3. Accessory minerals: Minerals with an average abundance of 1-2% in sedimentary rocks are referred as accessory minerals. These include common micas (muscovite, biotite), chlorite, clay minerals and heavy minerals which are denser than quartz. Average abundance of coarser mica in siliciclastic rocks is less than 0.5%. Muscovite is chemically more stable than biotite and also more abundant in sandstones. Micas are

generally derived from metamorphic source rocks, as well as, from plutonic igneous rocks.

Minerals having specific gravity more than 2.9 are categorised as heavy minerals. These include both chemically stable and unstable varieties. Zircon and rutile are widely studied heavy minerals and can survive multiple recycling episodes. If zircon and rutile are rounded then it indicates that last source was sedimentary. Relatively less stable minerals are magnetite, pyroxenes and amphiboles and less likely to survive recycling. These are generally first cycle sediments and reflect composition of proximate source rocks. Heavy minerals are derived from a variety of igneous, metamorphic and sedimentary rocks.

4. Rock fragments: Part of source rocks, yet not disintegrated, form individual grain in the rock are called rock fragments. Rock fragments, highly variable in composition, range from zero to sometimes 80% of rock. Fragments of any kind igneous, metamorphic or sedimentary rock may occur as rock fragments in sandstones. Most common rock fragments in sandstone are clasts of volcanic rock, volcanic glass, fine grained metamorphic rocks such as slate, phyllite, schist and quartzite. Sand size fragments of siltstone, shale, chert and chert nodules are also common. Clasts of limestone and other carbonate rocks are also found as rock fragments in sandstones. Rock fragments are particularly important to understand the sediment source rocks and more reliable indicators.

Chemical Minerals

Carbonate minerals in sandstones are both detrital and non-detrital (chemically precipitated). Detrital carbonate minerals include calcareous sand as skeletal fragments, oolites and faecal pellets mixed in minor amount in terrigenous sandstones. These components are not transported from a distant source.

Calcite and dolomite are commonly found in sandstones as pore filling and replacement cements of post-depositional origin. Additionally, Fe-Mn carbonates, sulfates and sulphides (pyrite, the most common) are found in sandstones as chemical minerals.

1. Mineral Cements: Framework grains in siliciclastic sedimentary rocks are bound together by some type of mineral cement. Cementing material may be silicate minerals such as quartz, opal or non-silicate minerals such as calcite or dolomite. Quartz is the most common mineral to act as cement in sandstones. In many ancient sandstones quartz cement is chemically attached to the crystal lattice of existing quartz grains forming rims of cement called quartz overgrowth. When silica cement is deposited as microcrystalline quartz, it fills interstitial spaces among framework grains. Carbonate minerals also commonly found as cement in siliciclastic sedimentary rocks. Calcite is most common and precipitates in the pore spaces of framework grains. Dolomite and siderite (iron carbonate) also found as cement. Other less commonly found cement in

sandstones includes hematite, limonite, feldspars, anhydrite, gypsum, barite, clay minerals and zeolite minerals. All cements are secondary minerals from after deposition of sandstones and during burial.

2. Matrix minerals: Grains smaller than 0.03mm which form interstitial spaces in sandstones are called matrix minerals? It may include clay minerals, fine size mica and feldspar. Clay minerals may be studied by X-ray diffraction and scanning electron microscopy. Most common clay minerals are illite, montmorillonite, kaolinite and chlorite. Clay minerals are formed during subaerial weathering and hydrolysis as secondary minerals and also by subaqueous weathering in marine environment and during diagenesis.

Chemical Composition of Sandstones

In coarse grained sediments identification of minerals give gross chemical composition but in fine grained rocks like shale, identification of minerals need special studies like X-ray diffractometer (XRD) analysis. Additionally, identification of trace elements and organic matter require separate studies like that of thermoluminescence. In the chemical analysis of sandstone SiO_2; TiO_2; Al_2O_3; Fe_2O_3; FeO; MgO; Cao; Na_2O; K_2O; P_2O_5 and S are found in variable proportions.

The average sedimentary rock contains about 58% SiO_2, 13% Al_2O_3, 6% CaO, 5% FeO and Fe_2O_3 in addition to minor amount of TiO_2, MgO, Na_2O, K_2O, P_2O_5, S and CO_2. Similarly, this composition suggests mixture of sandstone (mainly SiO_2), shale (mainly $Al_2O_3$and SiO_2) and limestone (mainly Cao and CO_2).

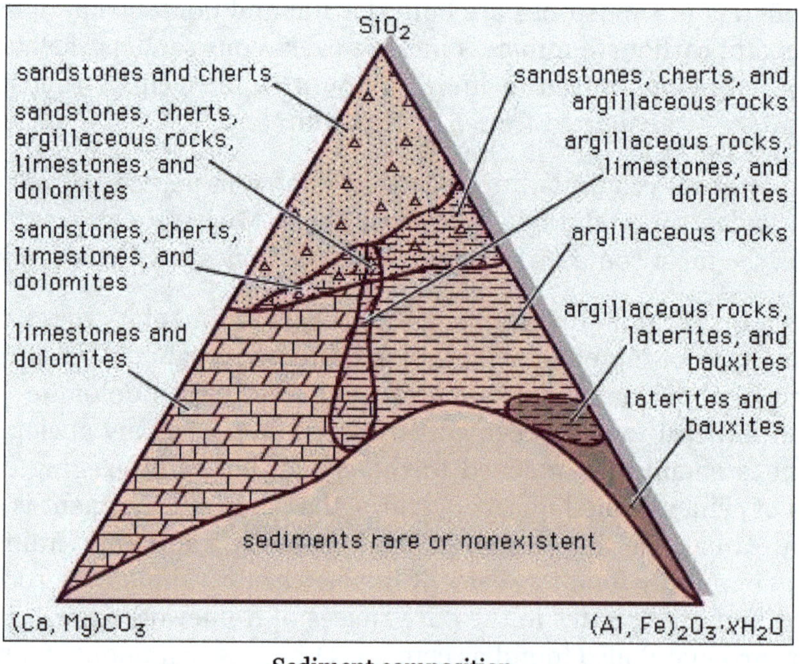

Sediment composition.

Texture and Structure

Texture

Texture is a fundamental attribute of siliciclastic sedimentary rocks and in sandstone it refers to the grain size, shape, roundness, surface features and arrangement of framework grains (fabric) of detrital components of sandstone. Other important part of texture study includes determination of porosity, permeability, bulk density, electrical conductivity, sound transmissibility and crushing strength. These derived properties are of particular interest to petroleum geologists, hydrologists, and geophysicists.

Permeability is the capacity of a rock to transmit fluid through it and its unit is darcy. This is measured in gallons per day through a one square foot cross section under a unit hydraulic gradient. Hydraulic gradient is the change in depth of the water table due to the direction of groundwater flow.

Textural maturity is defined as degree to which sand is free of interstitial clay and is well sorted and well rounded. These stages are based on the idea that during transportation clay is removed; framework grains are sorted and finally rounded in the process.

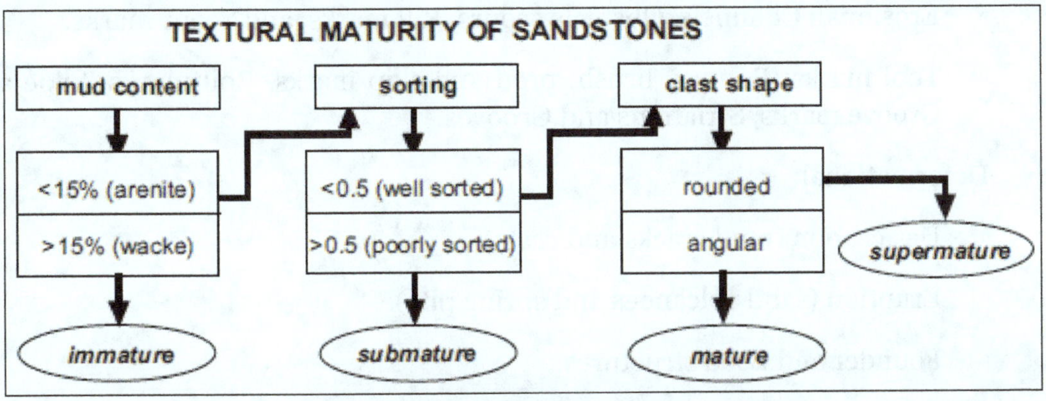

Textural maturity of sandstones.

Structure

Similar to texture and composition, sedimentary structures and bedding are inherent part of sandstones. Sedimentary structures and bedding both are identifiable by variations in grain size and mineralogy. Study of sedimentary structures is important because they are far and away the most valuable features for interpreting depositional environment.

Origins and geometries of sedimentary structures are highly varied. They can be classified on the basis of: kind of mechanism that produces them (physical sedimentary structures, chemical sedimentary structures, and biogenic sedimentary structures) and time of development relative to time of deposition (primary sedimentary structures and secondary sedimentary structures).

Table: Classification of sedimentary structures.

	Physical	Chemical	Biological
Primary	Stratification sole marks	Stratification	Biotubation (Tracks, Trails)
Secondary	Deformation Intrusion Desiccation	Nodules Concretions Deformation Stylolites	Bioturbation (Burrows)

Physical primary structures are certainly the most common and widespread. Most are related to transportation and deposition of sediment particles at a fluid/sediment interface. Such structures can be classified further on the basis of their relationship to transportation and deposition. Important sedimentary structures preserved in sandstones are:

- Current:
    - Depositional: Beach cusps, Graded bedding, Parallel lamination (Parting lineation), Ripple marks, Cross bedding, Wave and Swash marks.
    - Erosional: Channels, Obstacle scours, Rill marks, and Scour marks.
    - Tool marks: Bounce, brush, prod and skip marks, Roll marks, Slide and Groove marks, Striations and Grooves.
- Deformational:
    - Desiccation (mud cracks and casts).
    - Eruption (sand volcanoes and spring pits).
    - Founder and Load structures.
    - Impact (Spray, Hail, Rain pits).
    - Injection.
    - Slump.
- Biogenic:
    - Animal: Cawling trails, Feeding trails, Grazing trails, Resting trails.
    - Plant: Impression, Rootlets.
- Chemical:
    - Cementation (Sand crystals).
    - Crystallisation (Salt and Ice).

- Diffusion (Colour banding).
- Pressure solution (Stylolites).
- Replacement (Nodules).

Petrography

Using mineralogy and texture, petrography is the important tool to describe sandstones in meaningful way to have better understanding about origin and economic potential of sandstone. In all type of sands (terrigenous, carbonate and pyroclastic) relative proportions of framework grains, matrix and cement are the defining parameters for classification of sandstones. Framework grains in sandstones are quartz, feldspar and fine grained rock fragments (chert, limestone, siltstone, shale, slate, glass etc.). Mica are generally present but in subordinate amount as framework grains. Heavy minerals and glauconite are present in negligible amount.

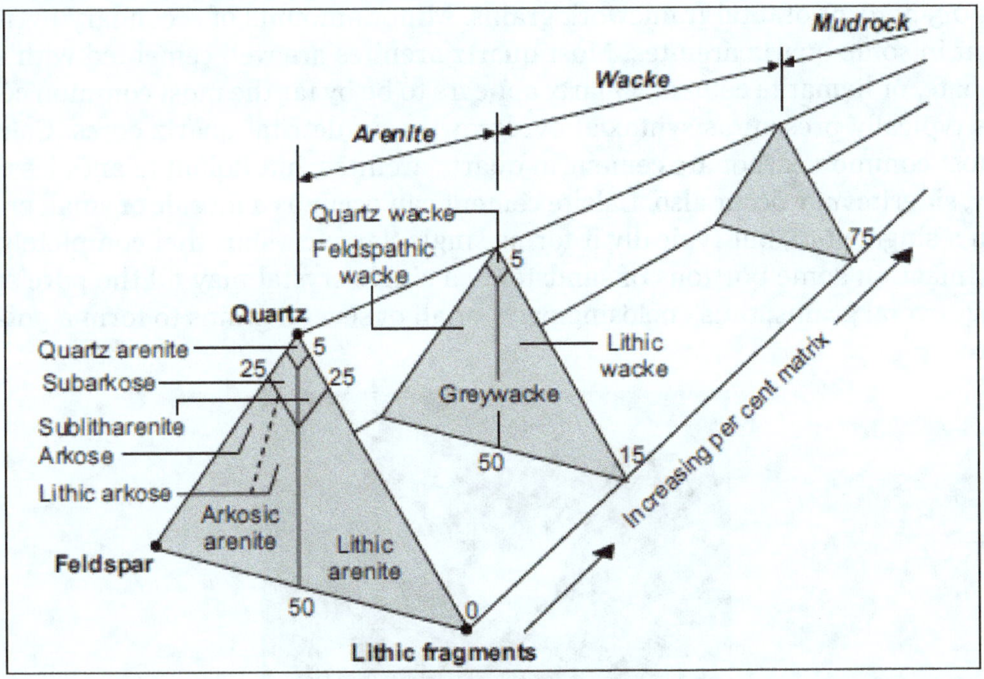

Classification of sandstones.

Quartz Arenites

Quartz arenites are composed of more than 90–95 percent siliceous grains (quartz, chert, quartzose rock fragments). They are generally well lithified and well cemented with silica or carbonate cement. Quartz arenites typically occur in association with assemblages of rocks deposited in stable cratonic environments such as eolian, beach, and shelf environments. Most quartz arenites are texturally mature to super mature. Cross-bedding is particularly characteristic of these rocks and ripple marks are also common. Siliceous grains in these sandstones range in abundance from 90 percent to more than 99 percent. Quartz dominates the siliceous constituents; however, a few percent chert, metaquartzite clasts, or siliceous sandstones/siltstone clasts may be included.

Quartz arenites tend to be characterized by high percentages of non-undulatory monocrystalline quartz, which may be due to relative enrichment in non-undulose quartz over time owing to selective destruction of less stable undulose quartz by mechanical and chemical processes operating during successive sedimentary cycles.

Feldspars make up trace amounts to as much as 5 percent of the framework constituents in many quartz arenites. Chert and silicic rock fragments, both metaquartzite clasts and sandstone/siltstone clasts, occur in some quartz arenites in amounts that rarely exceed l percent. Unstable rock fragments are very rare. Micas, commonly muscovite, occur in amounts that generally do not exceed about 0.5 percent. Heavy minerals are common in most quartzarenites, but generally make up no more than about 0.5 percent of total framework grains. Minor amounts of secondary pyrite are present in some quartz arenites. Most quartz arenites are well cemented with silica, carbonate, or hematite cement. Quartz appears to be by far the most common cement and is typically present as syntaxial overgrowths on detrital quartz cores. Calcite is the most common carbonate cement in quartz arenites, but dolomite and, less commonly, siderite may occur also. Calcite cement can occur as a mosaic of small crystals within a single pore, but typically it forms single, large crystals that completely fill a pore. Infact, in some portions of sandstone, a single crystal may fill the pore spaces among several grains, thus enclosing parts or all of several grains to form a poikilitic texture.

Typical quartz arenite.

Quartz arenites contain very less amount of matrix; many are matrix free. The small amount of matrix present in most quartz arenites is mainly due to diagenetic changes. The matrix consists mainly of fine-size quartz, feldspars, micas, and clay minerals, which likely form diagenetically by alteration of detrital feldspars and rock fragments.

Feldspathic Arenite

Feldspathic arenites contain less than 90 percent quartz grains, more feldspar than unstable rock fragments, and minor amounts of other minerals such as micas and heavy minerals. They may contain as little as 10 percent feldspar grains, but most feldspathic arenites show greater feldspar enrichment.

Quartz is the dominant particle constituent in most feldspathic arenites and typically makes about 50–60 percent of the framework grains. Quartz tends to be less abundant in plagioclase-rich feldspathic arenites than in K-feldspar-rich feldspathic arenites.

Carbonate cements appear to be the most common cements in feldspathic arenites and range from trace amounts to more than 20 percent of total constituents. Carbonate cements are typically calcite, but may include dolomite. Other cements, which generally occur in minor amounts, may include silica (quartz overgrowths), authigenic feldspars (overgrowths), hematite, and sulfate minerals such as barite, pyrite, and clay minerals.

The matrix content of feldspathic arenites may range from trace amounts to more than15 percent of total constituents. The matrix in many feldspathic arenites appears to consist mainly of sericite and kaolinite, typically stained with hematite. It is probably derived by alteration of biotite or other iron-bearing minerals. In others, the matrix materials may include also chlorite, fine quartz, feldspar, and organic matter.

Lithic Arenites

Lithic arenites are an extremely diverse group of rocks that are characterized by generally high content of unstable rock fragments. Any sandstone that contains less than 90 percent quartz (plus chert and quartzite) and unstable rock fragments in excess of feldspars is a lithic arenite. Many lithic arenites are poorly sorted; however, sorting ranges from well sorted to very poorly sorted. Quartz and many other framework grains are generally poorly rounded. Lithic arenites tend to contain substantial amounts of matrix, most of which may be of secondary origin.

Quartz Wacke

Sandstone with a large proportion (>15%) of fine grained matrix termed quartz wacke. Presence of large proportion of matrix, wide range of grain size, poor sorting of framework grains indicate textural immaturity. Large amount of feldspar and rock fragments show mineralogical immaturity. In the quartz wacke fine grained matrix is the main binding material and framework grains are generally not contact with each other. Large amount of matrix provide no scope for cement. Higher proportion of clay matrix indicate suspension load in transporting medium.

Arkose/Feldspathic Sandstone

Sandstone with high proportion (>10%) of feldspar is called feldspathic sandstone. When matrix content is also high then rock is called feldspathic wacke. Low matrix feldspathic sandstone are known as arkose or sub-arkose depending upon the proportion of feldspar. If feldspar is more than 25% then rock is arkose otherwise it is termed sub-arkose. Therefore, arkose refers to the sandstone with morethan 25% feldspar but less than 10% matrix content.

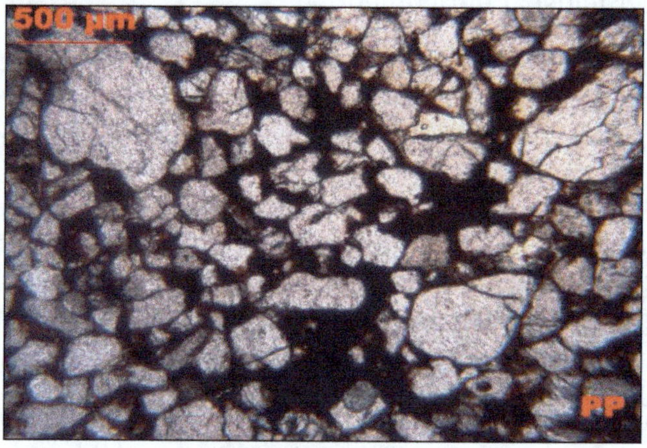

Typical quartz wacke.

Diagenesis

Diagenesis is the term used for all of the changes that sediment undergoes after deposition and before the transition to metamorphism. The multifarious processes that come under the term diagenesis are chemical, physical, and biological. They include compaction, deformation, dissolution, cementation, authigenesis, replacement, recrystallization, hydration, bacterial action, and development of concretions. Like other any other sedimentary rocks, diagenesis in sandstone is also an integral and important part. The two most important diagenetic processes are compaction and lithification. The study of diagenesis continues to be an active field of research in sedimentary geology, in part because the variety and complexity of the processes involved has left many uncertainties, but also because of the importance of diagenesis to petroleum geology, inasmuch as diagenesis is a significant control on porosity and permeability of deeply buried sedimentary rocks, comprising the fine siliciclastic source beds, the coarser reservoir rocks, and the seals that cause those reservoir rocks to be petroleum reservoirs. Diagenesis is an essential aspect, although quite different in processes and effects, of mud rocks, sandstones, and carbonate rocks.

Most sediment deposited under normal surface conditions has primary porosities of on the order of 30% to 70%. The lower values are more typical of coarser, sandy sediments, and the higher initial porosities are more typical of finer-grained, clay-rich sediments. Porosity in conglomerates and sandstones is largely a matter of pore spaces among the framework grains, as modified by later cementation together with certain other diagenetic changes.

Sandstones commonly show solution of grains at grain-to-grain contacts. This is especially important for quartz framework grains. This process is termed pressure solution. In its early stages, pressure solution involves only local areas at points of contact between adjacent grains. In later stages, a large percentage of the surface areas of adjacent grains are welded together. Such boundaries are said to be sutured. Thereby, almost all of the porosity of the rock has been eliminated.

As with most minerals, the solubility of quartz increases slightly with pressure. Upon substantial burial, local grain-to-grain pressures at grain contacts are enormous, not only because the weight of overlying rock is great but also because the surface areas of grain contacts is small. That leads to slight dissolution of quartz at the grain contacts. Consequently, the local concentration of silica in solution is slightly greater in the immediate vicinity of the contacts, creating a gradient in silica concentration away from the contacts. The surfaces of the quartz grains adjacent to the contacts, where pressures are much less, thus feel a slightly supersaturated pose solution, leading to precipitation of quartz on those surfaces. The action of this simultaneous dissolution and re-precipitation causes grain boundaries to become broader.

The diagenetic history of sandstone is controlled principally by the chemistry of the pore fluids that have moved through its pore system. The main factors that determine mineral precipitation or solution are- the chemistry of the sediment and the composition, concentration, Eh and pH of the pore fluids. Although many reactions occur during sandstone diagenesis, only a few are of major importance in sandstone cementation and porosity evolution.

Quartz-rich sandstones (quartz arenites) are commonly cemented to various degrees by silica. Usually this is quartz; only rarely is it in the form of opaline silica. Apart from pressure solution, development of secondary overgrowths of quartz upon quartz framework grains is another important aspect in quartz cementation. The new, authigenic quartz is precipitated in crystallographic continuity with the framework grains. The boundaries between the framework grains and the overgrowths are commonly marked by thin dusty-looking rims of very finegrained mineral material.

In sandstones with lower percentages of quartz framework grains (lithic arenites and feldspathic arenites) cementation by quartz is much less. In such sandstones and in many quartz arenites as well, the more common cementing material is calcite. In contrast to quartz cements, calcite cement tends to be patchy and irregular, suggesting replacement of earlier constituents of the sediment. Much of this replaced material is probably quartz itself as solubility of quartz increases with temperature whereas the solubility of calcite decreases with temperature. At sufficient depth of burial, quartz is dissolved and calcite is precipitated provided there is a sufficient concentration of calcium carbonate in the pore solutions passing through the rock.

Both the modes of cementation of a sandstone by quartz (pressure suturing and secondary overgrowths) result in loss of porosity, but pressure solution results in decrease of rock volume, whereas secondary overgrowths do not.

Economic Importance

Sedimentary rocks in general and sandstones in particular are pillars to sustain the modern world and critical to the mankind. Important mineral resources to sustain the economy and civilization are found in the sedimentary rocks. Sandstone makes the good aquifer for groundwater which is important source of water supply for various purposes. Similarly, sandstone is very good reservoir for hydrocarbons and throughout the world production of this natural resource is being taken out. Sandstone is also an important building material and host to many ores.

References

- Sedrx: tulane.edu, Retrieved 25, June 2020
- 8-biogenic, morelockonline, morelockSite: uprm.edu, Retrieved 06, April 2020
- Marine-Sediments-315191645: researchgate.net, Retrieved 12, January 2020

Chapter 3

Sedimentology Processes

Numerous processes are studied within the discipline of sedimentology. A few of them are sediment transport, deposition, erosion, diagenesis and weathering. The movement of organic and inorganic materials through the action of water or gravity is referred to as sediment transport. All these diverse sedimentology processes have been carefully analyzed in this chapter.

Sediment Transport

Sediment transport is the movement of solid particles (sediment), typically due to a combination of gravity acting on the sediment, and/or the movement of the fluid in which the sediment is entrained. Sediment transport occurs in natural systems where the particles are clastic rocks (sand, gravel, boulders, etc.), mud, or clay; the fluid is air, water, or ice; and the force of gravity acts to move the particles along the sloping surface on which they are resting. Sediment transport due to fluid motion occurs in rivers, oceans, lakes, seas, and other bodies of water due to currents and tides. Transport is also caused by glaciers as they flow, and on terrestrial surfaces under the influence of wind. Sediment transport due only to gravity can occur on sloping surfaces in general, including hillslopes, scarps, cliffs, and the continental shelf—continental slope boundary.

Dust blows from the Sahara Desert over the Atlantic Ocean towards the Canary Islands.

Sediment transport is important in the fields of sedimentary geology, geomorphology, civil engineering and environmental engineering. Knowledge of sediment transport is most often used to determine whether erosion or deposition will occur, the magnitude of this erosion or deposition, and the time and distance over which it will occur.

Mechanisms

Sand blowing off a crest in the Kelso Dunes of the Mojave Desert, California.

Toklat River, East Fork, Polychrome overlook, Denali National Park, Alaska. This river, like other braided streams, rapidly changes the positions of its channels through processes of erosion, sediment transport, and deposition.

Aeolian

Aeolian or *eolian* (depending on the parsing of æ) is the term for sediment transport by wind. This process results in the formation of ripples and sand dunes. Typically, the size of the transported sediment is fine sand (<1 mm) and smaller, because air is a fluid with low density and viscosity, and can therefore not exert very much shear on its bed.

Bedforms are generated by aeolian sediment transport in the terrestrial near-surface environment. Ripples and dunes form as a natural self-organizing response to sediment transport.

Aeolian sediment transport is common on beaches and in the arid regions of the world, because it is in these environments that vegetation does not prevent the presence and motion of fields of sand.

Wind-blown very fine-grained dust is capable of entering the upper atmosphere and moving across the globe. Dust from the Sahara deposits on the Canary Islands and islands in the Caribbean, and dust from the Gobi desert has deposited on the western United States. This sediment is important to the soil budget and ecology of several islands.

Deposits of fine-grained wind-blown glacial sediment are called loess.

Fluvial

In geology, physical geography, and sediment transport, fluvial processes relate to flowing water in natural systems. This encompasses rivers, streams, periglacial flows, flash floods and glacial lake outburst floods. Sediment moved by water can be larger than sediment moved by air because water has both a higher density and viscosity. In typical rivers the largest carried sediment is of sand and gravel size, but larger floods can carry cobbles and even boulders.

Fluvial sediment transport can result in the formation of ripples and dunes, in fractal-shaped patterns of erosion, in complex patterns of natural river systems, and in the development of floodplains.

Sand ripples, Laysan Beach, Hawaii. Coastal sediment transport results in these evenly spaced ripples along the shore. Monk seal for scale.

Coastal

Coastal sediment transport takes place in near-shore environments due to the motions of waves and currents. At the mouths of rivers, coastal sediment and fluvial sediment transport processes mesh to create river deltas.

Coastal sediment transport results in the formation of characteristic coastal landforms such as beaches, barrier islands, and capes.

A glacier joining the Gorner Glacier, Zermatt, Switzerland. These glaciers transport sediment and leave behind lateral moraines.

Glacial

As glaciers move over their beds, they entrain and move material of all sizes. Glaciers can carry the largest sediment, and areas of glacial deposition often contain a large number of glacial erratics, many of which are several metres in diameter. Glaciers also pulverize rock into "glacial flour", which is so fine that it is often carried away by winds to create loess deposits thousands of kilometres afield. Sediment entrained in glaciers often moves approximately along the glacial flowlines, causing it to appear at the surface in the ablation zone.

Hillslope

In hillslope sediment transport, a variety of processes move regolith downslope. These include:

- Soil creep
- Tree throw
- Movement of soil by burrowing animals
- Slumping and landsliding of the hillslope

These processes generally combine to give the hillslope a profile that looks like a solution to the diffusion equation, where the diffusivity is a parameter that relates to the ease of sediment transport on the particular hillslope. For this reason, the tops of hills generally have a parabolic concave-up profile, which grades into a convex-up profile around valleys.

As hillslopes steepen, however, they become more prone to episodic landslides and other mass wasting events. Therefore, hillslope processes are better described by a nonlinear diffusion equation in which classic diffusion dominates for shallow slopes and erosion rates go to infinity as the hillslope reaches a critical angle of repose.

Debris Flow

Large masses of material are moved in debris flows, hyperconcentrated mixtures of mud, clasts that range up to boulder-size, and water. Debris flows move as granular flows down steep mountain valleys and washes. Because they transport sediment as a granular mixture, their transport mechanisms and capacities scale differently from those of fluvial systems.

Applications

Suspended sediment from a stream emptying into a fjord (Isfjorden, Svalbard, Norway).

Sediment transport is applied to solve many environmental, geotechnical, and geological problems. Measuring or quantifying sediment transport or erosion is therefore important for coastal engineering. Several sediment erosion devices have been designed in order to quantitfy sediment erosion (e.g., Particle Erosion Simulator (PES)). One such device, also referred to as the BEAST (Benthic Environmental Assessment Sediment Tool) has been calibrated in order to quantify rates of sediment erosion.

Movement of sediment is important in providing habitat for fish and other organisms in rivers. Therefore, managers of highly regulated rivers, which are often sediment-starved due to dams, are often advised to stage short floods to refresh the bed material and rebuild bars. This is also important, for example, in the Grand Canyon of the Colorado River, to rebuild shoreline habitats also used as campsites.

Sediment discharge into a reservoir formed by a dam forms a reservoir delta. This delta will fill the basin, and eventually, either the reservoir will need to be dredged or the dam will need to be removed. Knowledge of sediment transport can be used to properly plan to extend the life of a dam.

Geologists can use inverse solutions of transport relationships to understand flow depth, velocity, and direction, from sedimentary rocks and young deposits of alluvial materials.

Flow in culverts, over dams, and around bridge piers can cause erosion of the bed. This erosion can damage the environment and expose or unsettle the foundations of the structure. Therefore, good knowledge of the mechanics of sediment transport in a built environment are important for civil and hydraulic engineers.

When suspended sediment transport is increased due to human activities, causing environmental problems including the filling of channels, it is called siltation after the grain-size fraction dominating the process.

Initiation of Motion

Stress Balance

For a fluid to begin transporting sediment that is currently at rest on a surface, the boundary (or bed) shear stress τ_b exerted by the fluid must exceed the critical shear stress τ_c for the initiation of motion of grains at the bed. This basic criterion for the initiation of motion can be written as:

$$\tau_b = \tau_c.$$

This is typically represented by a comparison between a dimensionless shear stress ($\tau_b *$)and a dimensionless critical shear stress ($\tau_c *$). The nondimensionalization is in order to compare the driving forces of particle motion (shear stress) to the resisting forces that would make it stationary (particle density and size). This dimensionless shear stress, $\tau*$, is called the Shields parameter and is defined as:

$$\tau* = \frac{\tau}{(\rho_s - \rho_f)(g)(D)}.$$

And the new equation to solve becomes:

$$\tau_b * = \tau_c *.$$

The equations included here describe sediment transport for clastic, or granular sediment. They do not work for clays and muds because these types of floccular sediments do not fit the geometric simplifications in these equations, and also interact thorough electrostatic forces. The equations were also designed for fluvial sediment transport of particles carried along in a liquid flow, such as that in a river, canal, or other open channel.

Only one size of particle is considered in this equation. However, river beds are often formed by a mixture of sediment of various sizes. In case of partial motion where only a part of the sediment mixture moves, the river bed becomes enriched in large gravel as the smaller sediments are washed away. The smaller sediments present under this layer of large gravel have a lower possibility of movement and total sediment transport decreases. This is called armouring effect. Other forms of armouring of sediment or decreasing rates of sediment erosion can be caused by carpets of microbial mats, under conditions of high organic loading.

Critical Shear Stress

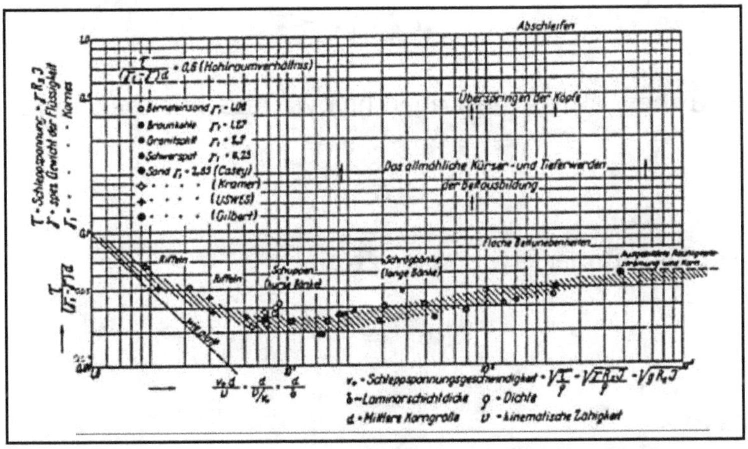

Original Shields diagram, 1936.

The Shields diagram empirically shows how the dimensionless critical shear stress (i.e. the dimensionless shear stress required for the initiation of motion) is a function of a particular form of the particle Reynolds number, Re_p or Reynolds number related to the particle. This allows us to rewrite the criterion for the initiation of motion in terms of only needing to solve for a specific version of the particle Reynolds number, which we call $\mathrm{Re}_p{}^*$.

$$\tau_b{}^* = f\left(\mathrm{Re}_p{}^*\right)$$

This equation can then be solved by using the empirically derived Shields curve to find $\tau_c{}^*$ as a function of a specific form of the particle Reynolds number called the boundary Reynolds number. The mathematical solution of the equation was given by Dey.

Particle Reynolds Number

In general, a particle Reynolds Number has the form:

$$\mathrm{Re}_p = \frac{U_p D}{\nu}$$

Where U_p is a characteristic particle velocity, D is the grain diameter (a characteristic particle size), and ν is the kinematic viscosity, which is given by the dynamic viscosity, μ, divided by the fluid density, ρ_f.

$$\nu = \frac{\mu}{\rho_f}$$

The specific particle Reynolds number of interest is called the boundary Reynolds number, and it is formed by replacing the velocity term in the Particle Reynolds number by the shear velocity, u_*, which is a way of rewriting shear stress in terms of velocity.

$$u_* = \sqrt{\frac{\tau_b}{\rho_f}} = \kappa z \frac{\partial u}{\partial z}$$

where τ_b is the bed shear stress (described below), and κ is the von Kármán constant, where

$$\kappa = 0.407.$$

The particle Reynolds number is therefore given by:

$$\mathrm{Re}_p{}^* = \frac{u_* D}{\nu}$$

Bed Shear Stress

The boundary Reynolds number can be used with the Shields diagram to empirically solve the equation

$$\tau_c{}^* = f\left(\mathrm{Re}_p{}^*\right),$$

which solves the right-hand side of the equation

$$\tau_b{}^* = \tau_c{}^*$$

In order to solve the left-hand side, expanded as

$$\tau_b{}^* = \frac{\tau_b}{(\rho_s - \rho_f)(g)(D)},$$

we must find the bed shear stress, τ_b. There are several ways to solve for the bed shear stress. First, we develop the simplest approach, in which the flow is assumed to be steady and uniform and reach-averaged depth and slope are used. Due to the difficulty of measuring shear stress *in situ*, this method is also one of the most-commonly used. This method is known as the depth-slope product.

Depth-slope Product

For a river undergoing approximately steady, uniform equilibrium flow, of approximately constant depth *h* and slope angle θ over the reach of interest, and whose width is much greater than its depth, the bed shear stress is given by some momentum considerations stating that the gravity force component in the flow direction equals exactly the friction force. For a wide channel, it yields:

$$\tau_b = \rho g h \sin(\theta)$$

For shallow slope angles, which are found in almost all natural lowland streams, the small-angle formula shows that $\sin(\theta)$ is approximately equal to $\tan(\theta)$, which is given by S, the slope. Rewritten with this:

$$\tau_b = \rho g h S$$

Shear Velocity, Velocity, and Friction Factor

For the steady case, by extrapolating the depth-slope product and the equation for shear velocity:

$$\tau_b = \rho g h S$$

$$u_* = \sqrt{\left(\frac{\tau_b}{\rho}\right)},$$

We can see that the depth-slope product can be rewritten as:

$$\tau_b = \rho u_*^2.$$

$u*$is related to the mean flow velocity, $\overline{u}$, through the generalized Darcy-Weisbach friction factor, C_f, which is equal to the Darcy-Weisbach friction factor divided by 8 (for mathematical convenience). Inserting this friction factor,

$$\tau_b = \rho C_f \left(\overline{u}\right)^2$$

Unsteady Flow

For all flows that cannot be simplified as a single-slope infinite channel (as in the depth-slope product, above), the bed shear stress can be locally found by applying the Saint-Vennant equations for continuity, which consider accelerations within the flow.

Example

Set-up

The criterion for the initiation of motion, established earlier, states that

$$\tau_b * = \tau_c *.$$

In this equation,

$$\tau * = \frac{\tau}{(\rho_s - \rho)(g)(D)},$$ and therefore

$$\frac{\tau_b}{(\rho_s - \rho)(g)(D)} = \frac{\tau_c}{(\rho_s - \rho)(g)(D)}.$$

τ_c * is a function of boundary Reynolds number, a specific type of particle Reynolds number.

$$\tau_c * = f\left(Re_p *\right).$$

For a particular particle Reynolds number, τ_c * will be an emprical constant given by the Shields Curve or by another set of empirical data (depending on whether or not the grain size is uniform).

Therefore, the final equation that we seek to solve is:

$$\frac{\tau_b}{(\rho_s - \rho)(g)(D)} = f\left(Re \ *\right).$$

Solution

We make several assumptions to provide an example that will allow us to bring the above form of the equation into a solved form.

First, we assume that the a good approximation of reach-averaged shear stress is given by the depth-slope product. We can then rewrite the equation as

$$\rho ghS = 0.06(\rho_s - \rho)(g)(D)..$$

Moving and re-combining the terms, we obtain:

$$hS = \frac{(\rho_s - \rho)}{\rho}(D)\left(f\left(\mathrm{Re}_p *\right)\right) = RD\left(f\left(\mathrm{Re}_p *\right)\right)$$

where R is the submerged specific gravity of the sediment.

We then make our second assumption, which is that the particle Reynolds number is high. This is typically applicable to particles of gravel-size or larger in a stream, and means that the critical shear stress is a constant. The Shields curve shows that for a bed with a uniform grain size,

$$\tau_c * = 0.06.$$

Later researchers have shown that this value is closer to

$$\tau_c * = 0.03$$

for more uniformly sorted beds. Therefore, we will simply insert

$$\tau_c * = f\left(\mathrm{Re}_p *\right)$$

and insert both values at the end.

The equation now reads:

$$hS = RD\tau_c{}^*$$

This final expression shows that the product of the channel depth and slope is equal to the Shield's criterion times the submerged specific gravity of the particles times the particle diameter.

For a typical situation, such as quartz-rich sediment $\left(\rho_s = 2650\frac{kg}{m^3}\right)$ in water $\left(\rho = 1000\frac{kg}{m^3}\right)$, the submerged specific gravity is equal to 1.65.

$$R = \frac{(\rho_s - \rho)}{\rho} = 1.65$$

Plugging this into the equation above,

$$hS = 1.65(D)\tau_c{}^*.$$

For the Shield's criterion of $\tau_c{}^* = 0.06.$. 0.06 * 1.65 = 0.099, which is well within standard margins of error of 0.1. Therefore, for a uniform bed,

$$hS = 0.1(D).$$

For these situations, the product of the depth and slope of the flow should be 10% of the diameter of the median grain diameter.

The mixed-grain-size bed value is $\tau_c{}^* = 0.03$, which is supported by more recent research as being more broadly applicable because most natural streams have mixed grain sizes. Using this value, and changing D to D_50 ("50" for the 50th percentile, or the median grain size, as we are now looking at a mixed-grain-size bed), the equation becomes:

$$hS = 0.05(D_{50})$$

Which means that the depth times the slope should be about 5% of the median grain diameter in the case of a mixed-grain-size bed.

Modes of Entrainment

The sediments entrained in a flow can be transported along the bed as bed load in the form of sliding and rolling grains, or in suspension as suspended load advected by the main flow. Some sediment materials may also come from the upstream reaches and be carried downstream in the form of wash load.

Rouse Number

The location in the flow in which a particle is entrained is determined by the Rouse number, which is determined by the density ρ_s and diameter d of the sediment particle, and the density ρ and kinematic viscosity v of the fluid, determine in which part of the flow the sediment particle will be carried.

$$P = \frac{w_s}{\kappa u_*}$$

Here, the Rouse number is given by P. The term in the numerator is the (downwards) sediment the sediment settling velocity w_s, which is discussed below. The upwards velocity on the grain is given as a product of the von Kármán constant, $\kappa = 0.4$, and the shear velocity, u_*.

The following table gives the approximate required Rouse numbers for transport as bed load, suspended load, and wash load.

Mode of Transport	Rouse Number
Initiation of motion	>7.5
Bed load	>2.5, <7.5
Suspended load: 50% Suspended	>1.2, <2.5
Suspended load: 100% Suspended	>0.8, <1.2
Wash load	<0.8

Settling Velocity

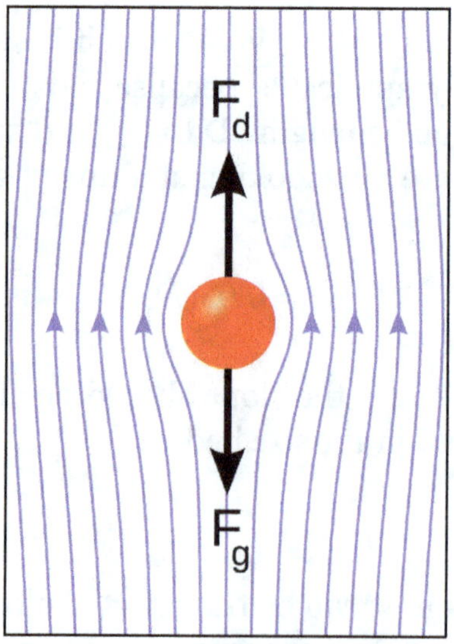

Streamlines around a sphere falling through a fluid. This illustration is accurate for laminar flow, in which the particle Reynolds number is small. This is typical for small particles falling through a viscous fluid; larger particles would result in the creation of a turbulent wake.

The settling velocity (also called the "fall velocity" or "terminal velocity") is a function of the particle Reynolds number. Generally, for small particles (laminar approximation), it can be calculated with Stokes' Law. For larger particles (turbulent particle Reynolds numbers), fall velocity is calculated with the turbulent drag law. Dietrich (1982) compiled a large amount of published data to which he empirically fit settling velocity curves. Ferguson and Church (2006) analytically combined the expressions for Stokes flow and a turbulent drag law into a single equation that works for all sizes of sediment, and successfully tested it against the data of Dietrich. Their equation is

$$w_s = \frac{RgD^2}{C_1\nu + (0.75C_2RgD^3)^{(0.5)}}.$$

In this equation w_s is the sediment settling velocity, g is acceleration due to gravity, and D is mean sediment diameter. ν is the kinematic viscosity of water, which is approximately 1.0 x 10^{-6} m^2/s for water at 20 °C.

C_1 and C_2 are constants related to the shape and smoothness of the grains.

Constant	Smooth Spheres	Natural Grains: Sieve Diameters	Natural Grains: Nominal Diameters	Limit for Ultra-Angular Grains
	18	18	20	24
	0.4	1.0	1.1	1.2

The expression for fall velocity can be simplified so that it can be solved only in terms of D. We use the sieve diameters for natural grains, $g = 9.8,,$ and values given above for ν and R. From these parameters, the fall velocity is given by the expression:

$$w_s = \frac{16.17D^2}{1.8\cdot10^{-5} + (12.1275D^3)^{(0.5)}}$$

Hjulström-Sundborg Diagram

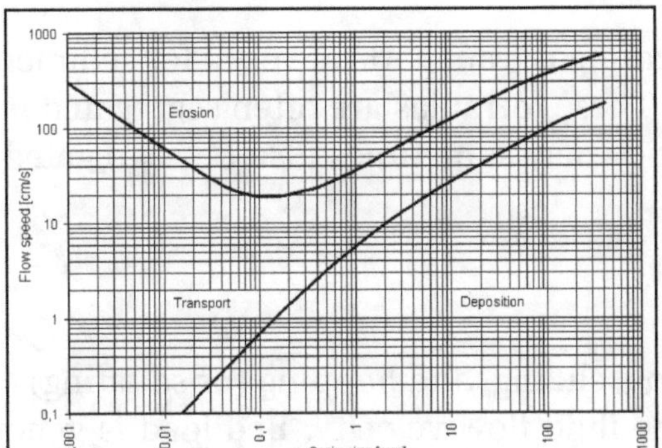

The logarithmic Hjulström curve.

In 1935, Filip Hjulström created the Hjulström curve, a graph which shows the relationship between the size of sediment and the velocity required to erode (lift it), transport it, or deposit it. The graph is logarithmic.

Åke Sundborg later modified the Hjulström curve to show separate curves for the movement threshold corresponding to several water depths, as is necessary if the flow velocity rather than the boundary shear stress (as in the Shields diagram) is used for the flow strength.

This curve has no more than a historical value nowadays, although its simplicity is still attractive. Among the drawbacks of this curve are that it does not take the water depth into account and more importantly, that it does not show that sedimentation is caused by flow velocity *deceleration* and erosion is caused by flow *acceleration*. The dimensionless Shields diagram is now unanimously accepted for initiation of sediment motion in rivers. Much work was done on river sediment transport formulae in the second half of the 20th century and that work should be used preferably to Hjulström's curve, e.g. Meyer-Peter & Müller (1948), Engelund-Hansen (1967), Lefort (1991), Belleudy (2012).

Transport Rate

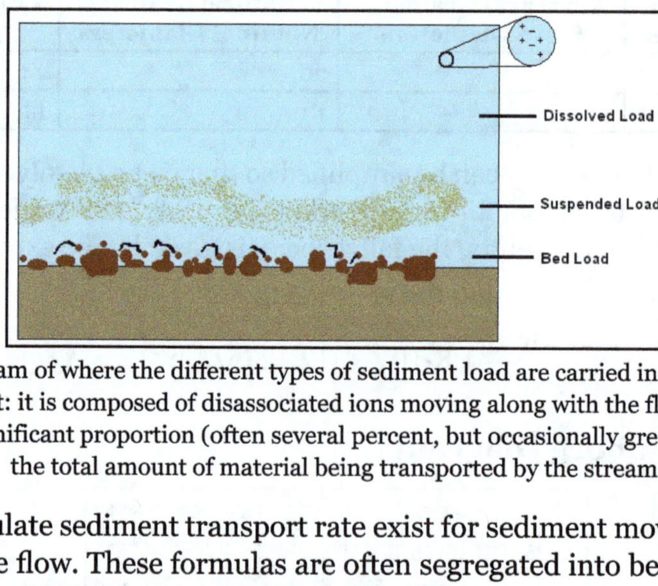

A schematic diagram of where the different types of sediment load are carried in the flow. Dissolved load is not sediment: it is composed of disassociated ions moving along with the flow. It may, however, constitute a significant proportion (often several percent, but occasionally greater than half) of the total amount of material being transported by the stream.

Formulas to calculate sediment transport rate exist for sediment moving in several different parts of the flow. These formulas are often segregated into bed load, suspended load, and wash load. They may sometimes also be segregated into bed material load and wash load.

Bed Load

Bed load moves by rolling, sliding, and hopping (or saltating) over the bed, and moves at a small fraction of the fluid flow velocity. Bed load is generally thought to constitute 5-10% of the total sediment load in a stream, making it less important in terms of mass balance. However, the bed material load (the bed load plus the portion of the suspended load which comprises material derived from the bed) is often dominated by bed load, especially in gravel-bed rivers. This bed material load is the only part of the sediment load that actively interacts with the bed. As the bed load is an important component of that, it plays a major role in controlling the morphology of the channel.

Bed load transport rates are usually expressed as being related to excess dimensionless shear stress raised to some power. Excess dimensionless shear stress is a nondimensional measure of bed shear stress about the threshold for motion.

$$(\tau_b^* - \tau_c^*),$$

Bed load transport rates may also be given by a ratio of bed shear stress to critical shear stress, which is equivalent in both the dimensional and nondimensional cases. This ratio is called the "transport stage" (T_s or ϕ) and is an important in that it shows bed shear stress as a multiple of the value of the criterion for the initiation of motion.

$$T_s = \phi = \frac{\tau_b}{\tau_c}$$

When used for sediment transport formulae, this ratio is typically raised to a power.

The majority of the published relations for bedload transport are given in dry sediment weight per unit channel width, b ("breadth"):

$$q_s = \frac{Q_s}{b}.$$

Due to the difficulty of estimating bed load transport rates, these equations are typically only suitable for the situations for which they were designed.

Notable Bed Load Transport Formulae

Meyer-Peter Müller and Derivatives

The transport formula of Meyer-Peter and Müller, originally developed in 1948, was designed for well-sorted fine gravel at a transport stage of about 8. The formula uses the above nondimensionalization for shear stress,

$$\tau^* = \frac{\tau}{(\rho_s - \rho)(g)(D)},$$

and Hans Einstein's nondimensionalization for sediment volumetric discharge per unit width

$$q_s^* = \frac{q_s}{D\sqrt{\frac{\rho_s - \rho}{\rho} gD}} = \frac{q_s}{Re_p \nu}.$$

Their formula reads:

$$q_s^* = 8\left(\tau^* - \tau^*_c\right)^{3/2}.$$

Their experimentally determined value for τ^*_c is 0.047, and is the third commonly used value for this (in addition to Parker's 0.03 and Shields' 0.06).

Because of its broad use, some revisions to the formula have taken place over the years that show that the coefficient on the left ("8" above) is a function of the transport stage:

$$T_s \approx 2 \rightarrow q_s^* = 5.7\left(\tau^* - 0.047\right)^{3/2}$$

$$T_s \approx 100 \rightarrow q_s^* = 12.1\left(\tau^* - 0.047\right)^{3/2}$$

The variations in the coefficient were later generalized as a function of dimensionless shear stress:

$$\begin{cases} q_s^* = \alpha_s\left(\tau^* - \tau_c^*\right)^n \\ n = \dfrac{3}{2} \\ \alpha_s = 1.6\ln\left(\tau^*\right) + 9.8 \approx 9.64\tau^{*0.166} \end{cases}$$

Wilcock and Crowe

In 2003, Peter Wilcock and Joanna Crowe (now Joanna Curran) published a sediment transport formula that works with multiple grain sizes across the sand and gravel range. Their formula works with surface grain size distributions, as opposed to older models which use subsurface grain size distributions (and thereby implicitly infer a surface grain sorting).

Their expression is more complicated than the basic sediment transport rules (such as that of Meyer-Peter and Müller) because it takes into account multiple grain sizes: this requires consideration of reference shear stresses for each grain size, the fraction of the total sediment supply that falls into each grain size class, and a "hiding function".

The "hiding function" takes into account the fact that, while small grains are inherently more mobile than large grains, on a mixed-grain-size bed, they may be trapped in deep pockets between large grains. Likewise, a large grain on a bed of small particles will be stuck in a much smaller pocket than if it were on a bed of grains of the same size. In gravel-bed rivers, this can cause "equal mobility", in which small grains can move just as easily as large ones. As sand is added to the system, it moves away from the "equal mobility" portion of the hiding function to one in which grain size again matters.

Their model is based on the transport stage, or ratio of bed shear stress to critical shear stress for the initiation of grain motion. Because their formula works with several grain sizes simultaneously, they define the critical shear stress for each grain size class, τ_{c,D_i}, to be equal to a "reference shear stress", τ_{ri}. .

They express their equations in terms of a dimensionless transport parameter, W_i^* (with the "$*$" indicating nondimensionality and the "i" indicating that it is a function of grain size):

$$W_i^* = \frac{Rgq_{bi}}{F_i u*^3}$$

q_{bi} is the volumetric bed load transport rate of size class i per unit channel width b. F_i is the proportion of size class i that is present on the bed.

They came up with two equations, depending on the transport stage, ϕ. For $\phi < 1.35$:

$$W_i^* = 0.002\phi^{7.5}$$

and for $\phi \geq 1.35$:

$$W_i^* = 14\left(1 - \frac{0.894}{\phi^{0.5}}\right)^{4.5}.$$

This equation asymptotically reaches a constant value of W_i^* as ϕ becomes large.

Wilcock and Kenworthy

In 2002, Peter Wilcock and Kenworthy T.A. , following Peter Wilcock (1998), published a sediment bed-load transport formula that works with only two sediments fractions, i.e. sand and gravel fractions. Peter Wilcock and Kenworthy T.A. in their article recognized that a mixed-sized sediment bed-load transport model using only two fractions offers practical advantages in terms of both computational and conceptual modeling by taking into account the nonlinear effects of sand presence in gravel beds on bed-load transport rate of both fractions. In fact, in the two-fraction bed load formula appears a new ingredient with respect to that of Meyer-Peter and Müller that is the proportion F_i of fraction i on the bed surface where the subscript $_i$ represents either the sand (s) or gravel (g) fraction. The proportion F_i, as a function of sand content f_s, physically represents the relative influence of the mechanisms controlling sand and gravel transport, associated with the change from a clast-supported to matrix-supported gravel bed. Moreover, since f_s spans between 0 and 1, phenomena that vary with f_s include the relative size effects producing "hiding" of fine grains and "exposure" of coarse grains. The "hiding" effect takes into account the fact that, while small grains are inherently more mobile than large grains, on a mixed-grain-size bed, they may be trapped in deep pockets between large grains. Likewise, a large grain on a bed of small particles will be stuck in a much smaller pocket than if it were on a bed of grains of the same size, which the Meyer-Peter and Müller formula refers to. In gravel-bed rivers, this can cause "equal mobility", in which small grains can move just as easily as large ones. As sand is added to the system, it moves away from the "equal mobility" portion of the hiding function to one in which grain size again matters.

Their model is based on the transport stage,*i.e.* ϕ, or ratio of bed shear stress to critical shear stress for the initiation of grain motion. Because their formula works with only two fractions simultaneously, they define the critical shear stress for each of the two grain size classes, τ_{ri}, where $_i$ represents either the sand (s) or gravel (g) fraction . The critical shear stress that represents the incipient motion for each of the two fractions is consistent with established values in the limit of pure sand and gravel beds and shows a sharp change with increasing sand content over the transition from a clast- to matrix-supported bed.

They express their equations in terms of a dimensionless transport parameter, W_i^* (with the "$*$" indicating nondimensionality and the "$_i$" indicating that it is a function of grain size):

$$W_i^* = \frac{Rgq_{bi}}{F_i u^{*3}}$$

q_{bi} is the volumetric bed load transport rate of size class i per unit channel width b . F_i is the proportion of size class i that is present on the bed.

They came up with two equations, depending on the transport stage, ϕ. For $\phi < \phi$:

$$W_i^* = 0.002\phi^{7.5}$$

and for $\phi \geq \phi$:

$$W_i^* = A\left(1 - \frac{\chi}{\phi^{0.5}}\right)^{4.5}$$

This equation asymptotically reaches a constant value of W_i^* as ϕ becomes large and the symbols A, ϕ', χ have the following values:

$$A = 70, \phi' = 1.19, \chi = 0.908, laboratory$$

$$A = 115, \phi' = 1.27, \chi = 0.923, field$$

In order to apply the above formulation, it is necessary to specify the characteristic grain sizes D_s for the sand portion and D_g for the gravel portion of the surface layer, the fractions F_s and F_g of sand and gravel, respectively in the surface layer, the submerged specific gravity of the sediment R and shear velocity associated with skin friction u_*.

Kuhnle et al.

For the case in which sand fraction is transported by the current over and through an immobile gravel bed, Kuhnle *et al.*(2013), following the theoretical analysis done by Pellachini (2011), provides a new relationship for the bed load transport of the sand fraction when gravel particles remain at rest. It is worth mentioning that Kuhnle *et al.* (2013) applied the Wilcock and Kenworthy (2002) formula to their experimental data and found out that predicted bed load rates of sand fraction were about 10 times greater than measured and approached 1 as the sand elevation became near the top of the gravel layer. They, also, hypothesized that the mismatch between predicted and measured sand bed load rates is due to the fact that the bed shear stress used for the Wilcock and Kenworthy (2002) formula was larger than that available for transport within the gravel bed because of the sheltering effect of the gravel particles. To overcome this mismatch, following Pellachini (2011), they assumed that the variability of the bed shear stress available for the sand to be transported by the current would be some function of the so-called "Roughness Geometry Function" (RGF), which represents the gravel bed elevations distribution. Therefore, the sand bed load formula follows as:

$$q_s^* = 2.29 * 10^{-5} A(z_s)^{2.14} \left(\frac{\tau_b}{\tau_{cs}} \right)^{3.49}$$

where

$$q_s^* = \frac{q_s}{[(s-1)gD_s]^{0.5} \rho_s D_s}$$

the subscript $_s$ refers to the sand fraction, s represents the ratio ρ_s / ρ_w where ρ_s is the sand fraction density, $A(z_s)$ is the RGF as a function of the sand level z_s within the gravel bed, τ_b is the bed shear stress available for sand transport and τ_{cs} is the critical shear stress for incipient motion of the sand fraction, which was calculated graphically using the updated Shields-type relation of Miller *et al.*(1977).

Suspended Load

Suspended load is carried in the lower to middle parts of the flow, and moves at a large fraction of the mean flow velocity in the stream.

A common characterization of suspended sediment concentration in a flow is given by the Rouse Profile. This characterization works for the situation in which sediment concentration c_0 at one particular elevation above the bed z_0 can be quantified. It is given by the expression:

$$\frac{c_s}{c_0} = \left[\frac{z(h - z_0)}{z_0(h - z)} \right]^{-P/\alpha}$$

Here, z is the elevation above the bed, c_s is the concentration of suspended sediment at that elevation, h is the flow depth, P is the Rouse number, and α relates the eddy viscosity for momentum K_m to the eddy diffusivity for sediment, which is approximately equal to one.

$$\alpha = \frac{K_s}{K_m} \approx 1$$

Experimental work has shown that α ranges from 0.93 to 1.10 for sands and silts.

The Rouse profile characterizes sediment concentrations because the Rouse number includes both turbulent mixing and settling under the weight of the particles. Turbulent mixing results in the net motion of particles from regions of high concentrations to low concentrations. Because particles settle downward, for all cases where the particles are not neutrally buoyant or sufficiently light that this settling velocity is negligible, there is a net negative concentration gradient as one goes upward in the flow. The Rouse Profile therefore gives the concentration profile that provides a balance between turbulent mixing (net upwards) of sediment and the downwards settling velocity of each particle.

Bed Material Load

Bed material load comprises the bed load and the portion of the suspended load that is sourced from the bed.

Three common bed material transport relations are the "Ackers-White", "Engelund-Hansen", "Yang" formulae. The first is for sand to granule-size gravel, and the second and third are for sand though Yang later expanded his formula to include fine gravel. That all of these formulae cover the sand-size range and two of them are exclusively for sand is that the sediment in sand-bed rivers is commonly moved simultaneously as bed and suspended load.

Engelund-Hansen

The bed material load formula of Engelund and Hansen is the only one to not include some kind of critical value for the initiation of sediment transport. It reads:

$$q_s{*} = \frac{0.05}{c_f} \tau{*}^{2.5}$$

where $q_s{*}$ is the Einstein nondimensionalization for sediment volumetric discharge per unit width, c_f is a friction factor, and $\tau{*}$ is the Shields stress. The Engelund-Hansen formula is one of the few sediment transport formulae in which a threshold "critical shear stress" is absent.

Wash Load

Wash load is carried within the water column as part of the flow, and therefore moves with the mean velocity of main stream. Wash load concentrations are approximately uniform in the water column. This is described by the endmember case in which the Rouse number is equal to 0 (i.e. the settling velocity is far less than the turbulent mixing velocity), which leads to a prediction of a perfectly uniform vertical concentration profile of material.

Total Load

Some authors have attempted formulations for the total sediment load carried in water. These formulas are designed largely for sand, as (depending on flow conditions) sand often can be carried as both bed load and suspended load in the same stream or shoreface.

Deposition

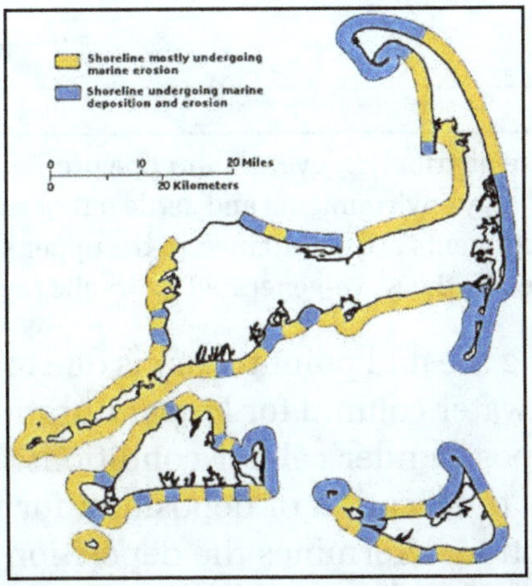

Map of Cape Cod showing shores undergoing erosion (cliffed sections) in yellow, and shores characterized by marine deposition (barriers) in blue.

Deposition is the geological process in which sediments, soil and rocks are added to a landform or land mass. Wind, ice, and water, as well as sediment flowing via gravity, transport previously eroded sediment, which, at the loss of enough kinetic energy in the fluid, is deposited, building up layers of sediment.

Deposition occurs when the forces responsible for sediment transportation are no longer sufficient to overcome the forces of gravity and friction, creating a resistance to motion, this is known as the null-point hypothesis. Deposition can also refer to the buildup of sediment from organically derived matter or chemical processes. For example, chalk is made up partly of the microscopic calcium carbonate skeletons of marine plankton, the deposition of which has induced chemical processes (diagenesis) to deposit further calcium carbonate. Similarly, the formation of coal begins with deposition of organic material, mainly from plants, in anaerobic conditions.

Null-point Hypothesis

The null-point hypothesis explains how sediment is deposited throughout a shore profile according to its grain size. This is due to the influence of hydraulic energy, resulting in a seaward-fining of sediment particle size, or where fluid forcing equals gravity for each grain size. The concept can also be explained as "sediment of a particular size may move across the profile to a position where it is in equilibrium with the wave and flows acting on that sediment grain". This sorting mechanism combines the influence of the down-slope gravitational force of the profile and forces due to flow asymmetry, the position where there is zero net transport is known as the null point and was first proposed by Cornaglia in 1889. Figure 1 illustrates this relationship between sediment grain size and the depth of the marine environment.

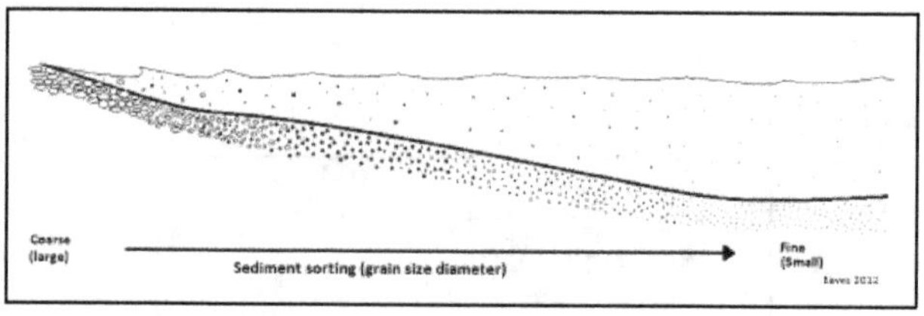

Illustrates the sediment size distribution over a shoreline profile, where finer sediments are transported away from high energy environments and settle out of suspension, or deposit in calmer environments. Coarse sediments are maintained in the upper shoreline profile and are sorted by the wave-generated hydraulic regime.

The first principle underlying the null point theory is due to the gravitational force; finer sediments remain in the water column for longer durations allowing transportation outside the surf zone to deposit under calmer conditions. The gravitational effect, or settling velocity determines the location of deposition for finer sediments, whereas a grain's internal angle of friction determines the deposition of larger grains on a shore profile. The secondary principle to the creation of seaward sediment fining is known as the hypothesis of asymmetrical thresholds under waves; this describes the interaction between the oscillatory flow of waves and tides flowing over the wave ripple bedforms in an asymmetric pattern. "The relatively strong onshore stroke of the wave forms an eddy or vortex on the lee side of the ripple, provided the onshore flow persists, this eddy remains trapped in the lee of the ripple. When the flow reverses, the eddy is thrown upwards off the bottom and a small cloud of suspended sediment generated by the eddy is ejected into the water column above the ripple, the sediment cloud is then moved seaward by the offshore stroke of the wave." Where there is symmetry in ripple shape the vortex is neutralised, the eddy and its associated sediment cloud develops on both sides of the ripple. This creates a cloudy water column which travels under tidal influence as the wave orbital motion is in equilibrium.

The Null-point hypothesis has been quantitatively proven in Akaroa Harbour, New Zealand, The Wash, U.K., Bohai Bay and West Huang Sera, Mainland China, and in numerous other studies; Ippen and Eagleson (1955), Eagleson and Dean (1959, 1961) and Miller and Zeigler (1958, 1964).

Deposition of Non-cohesive Sediments

Large grain sediments transported by either bed load or suspended load will come to rest when there is insufficient bed shear stress and fluid turbulence to keep the sediment moving, with the suspended load this can be some distance as the particles need to fall through the water column. This is determined by the grains downward acting weight force being matched by a combined buoyancy and fluid drag force and can be expressed by:

$$\frac{4}{3}\pi R^3 \rho_s g = \frac{4}{3}\pi R^3 \rho g + \frac{1}{2} C_d \rho \pi R^2 w_s^2$$

Downward acting weight force = Upward-acting buoyancy force + Upward-acting fluid drag force

where:

- π is the ratio of a circle's circumference to its diameter.
- R is the radius of the spherical object (in m),
- ρ is the mass density of the fluid (kg/m^3),
- g is the gravitational acceleration (m/s^2),
- C_d is the drag coefficient, and
- w_s is the particle's settling velocity (in m/s).

In order to calculate the drag coefficient, the grain's Reynolds number needs to be discovered, which is based on the type of fluid through which the sediment particle is flowing; laminar flow, turbulent flow or a hybrid of both. When the fluid becomes more viscous due to smaller grain sizes or larger settling velocities, prediction is less straight forward and it is applicable to incorporate Stokes Law (also known as the frictional force, or drag force) of settling.

Deposition of Cohesive Sediments

Cohesion of sediment occurs with the small grain sizes associated with silts and clays, or particles smaller than 4□ on the phi scale. If these fine particles remain dispersed in the water column, Stokes law applies to the settling velocity of the individual grains, although due to sea water being a strong electrolyte bonding agent, flocculation occurs where individual particles create an electrical bond adhering each other together to form flocs. "The face of a clay platelet has a slight negative charge where the edge has a slight positive charge, when two platelets come into close proximity with each other the face of one particle and the edge of the other are electrostatically attracted." Flocs then have a higher combined mass which leads to quicker deposition through a higher fall velocity, and deposition in a more shoreward direction than they would have as the individual fine grains of clay or silt.

The Occurrence of Null Point Theory

Akaroa Harbour is located on Banks Peninsula, Canterbury, New Zealand,

43°48′S 172°56′E43.800°S 172.933°E. The formation of this harbour has occurred due to active erosional processes on an extinct shield volcano, whereby the sea has flooded the caldera creating an inlet 16 km in length, with an average width of 2 km and a depth of -13 m relative to mean sea level at the 9 km point down the transect of the central axis. The predominant storm wave energy has unlimited fetch for the outer harbour from a southerly direction, with a calmer environment within the inner harbour, though localised harbour breezes create surface currents and chop influencing the marine sedimentation processes. Deposits of loess from subsequent glacial periods have in filled volcanic fissures over millennia, resulting in volcanic basalt and loess as the main sediment types available for deposition in Akaroa Harbour

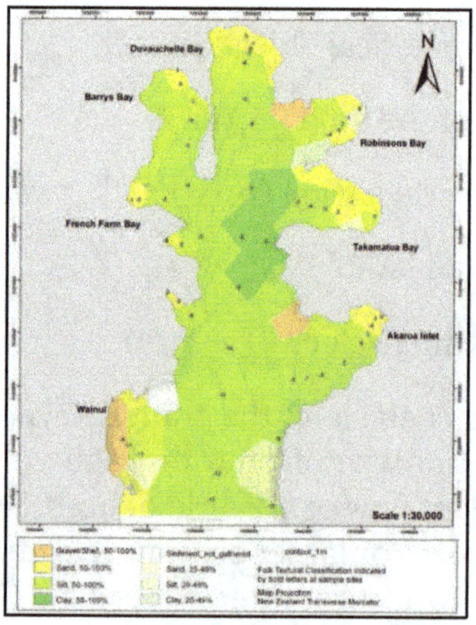

Figure. Map of Akaroa Harbour showing a fining of sediments with increased bathymetry toward the central axis of the harbour. Taken from Hart et al. (2009) and the University of Canterbury under contract of Environment Canterbury.

Hart et al. (2009) discovered through bathymetric survey, sieve and pipette analysis of subtidal sediments, that sediment textures were related to three main factors: depth; distance from shoreline; and distance along the central axis of the harbour. Resulting in the fining of sediment textures with increasing depth and towards the central axis of the harbour, or if classified into grain class sizes, "the plotted transect for the central axis goes from silty sands in the intertidal zone, to sandy silts in the inner nearshore, to silts in the outer reaches of the bays to mud at depths of 6 m or more"

Other studies have shown this process of the winnowing of sediment grain size from the effect of hydrodynamic forcing; Wang, Collins and Zhu (1988) qualitatively correlated increasing intensity of fluid forcing with increasing grain size. "This correlation was demonstrated at the low energy clayey tidal flats of Bohai Bay (China), the moderate environment of the Jiangsu coast (China) where the bottom material is silty, and the sandy flats of the high energy coast of The Wash (U.K.)." This research shows conclusive evidence for the null point theory existing on tidal flats with differing hydrodynamic energy levels and also on flats that are both erosional and accretional.

Kirby R. (2002) takes this concept further explaining that the fines are suspended and reworked aerially offshore leaving behind lag deposits of mainly bivalve and gastropod shells separated out from the finer substrate beneath, waves and currents then heap these deposits to form chenier ridges throughout the tidal zone which tend to be forced up the foreshore profile but also along the foreshore. Cheniers can be found at any level on the foreshore and predominantly characterise an erosion-dominated regime.

Applications for Coastal Planning and Management

The null point theory has been controversial in its acceptance into mainstream coastal science as the theory operates in dynamic equilibrium or unstable equilibrium, and many field and laboratory observations have failed to replicate the state of a null point at each grain size throughout the profile. The interaction of variables and processes over time within the environmental context causes issues; "the large number of variables, the complexity of the processes, and the difficulty in observation, all place serious obstacles in the way of systematisation, therefore in certain narrow fields the basic physical theory may be sound and reliable but the gaps are large"

Geomorphologists, engineers, governments and planners should be aware of the processes and outcomes involved with the null point hypothesis when performing tasks such as beach nourishment, issuing building consents or building coastal defence structures. This is because sediment grain size analysis throughout a profile allows inference into the erosion or accretion rates possible if shore dynamics are modified. Planners and managers should also be aware that the coastal environment is dynamic and contextual science should be evaluated before implementation of any shore profile modification. Thus theoretical studies, laboratory experiments, numerical and hydraulic modelling seek to answer questions pertaining to littoral drift and sediment deposition, the results should not be viewed in isolation and a substantial body of purely qualitative observational data should supplement any planning or management decision.

Erosion

An actively eroding rill on an intensively-farmed field in eastern Germany.

In earth science, erosion is the action of surface processes (such as water flow or wind) that remove soil, rock, or dissolved material from one location on the Earth's crust, then transport it away to another location. The particulate breakdown of rock or soil into clastic sediment is referred to as *physical* or *mechanical* erosion; this contrasts with *chemical* erosion, where soil or rock material is removed from an area by its dissolving into a solvent (typically water), followed by the flow away of that solution. Eroded sediment or solutes may be transported just a few millimetres, or for thousands of kilometres.

Natural rates of erosion are controlled by the action of geomorphic drivers, such as rainfall; bedrock wear in rivers; coastal erosion by the sea and waves; glacial plucking, abrasion, and scour; areal flooding; wind abrasion; groundwater processes; and mass movement processes in steep landscapes like landslides and debris flows. The rates at which such processes act control how fast a surface is eroded. Typically, physical erosion proceeds fastest on steeply sloping surfaces, and rates may also be sensitive to some climatically-controlled properties including amounts of water supplied (e.g., by rain), storminess, wind speed, wave fetch, or atmospheric temperature (especially for some ice-related processes). Feedbacks are also possible between rates of erosion and the amount of eroded material that is already carried by, for example, a river or glacier. Processes of erosion that produce sediment or solutes from a place contrast with those of deposition, which control the arrival and emplacement of material at a new location.

While erosion is a natural process, human activities have increased by 10-40 times the rate at which erosion is occurring globally. Excessive (or accelerated) erosion causes both "on-site" and "off-site" problems. On-site impacts include decreases in agricultural productivity and (on natural landscapes) ecological collapse, both because of loss of the nutrient-rich upper soil layers. In some cases, the eventual end result is desertification. Off-site effects include sedimentation of waterways and eutrophication of water bodies, as well as sediment-related damage to roads and houses. Water and wind erosion are the two primary causes of land degradation; combined, they are responsible for about 84% of the global extent of degraded land, making excessive erosion one of the most significant environmental problems worldwide.

Intensive agriculture, deforestation, roads, anthropogenic climate change and urban sprawl are amongst the most significant human activities in regard to their effect on stimulating erosion. However, there are many prevention and remediation practices that can curtail or limit erosion of vulnerable soils.

A natural arch produced by the wind erosion of differentially weathered rock in Jebel Kharaz, Jordan.

A wave-like sea cliff produced by coastal erosion, in Jinshitan Coastal National Geopark, Dalian, Liaoning Province, China.

Physical Processes

Rainfall and Surface Runoff

Soil and water being splashed by the impact of a single raindrop.

Rainfall, and the surface runoff which may result from rainfall, produces four main types of soil erosion: *splash erosion*, *sheet erosion*, *rill erosion*, and *gully erosion*. Splash erosion is generally seen as the first and least severe stage in the soil erosion process, which is followed by sheet erosion, then rill erosion and finally gully erosion (the most severe of the four).

In *splash erosion*, the impact of a falling raindrop creates a small crater in the soil, ejecting soil particles. The distance these soil particles travel can be as much as 0.6 m (two feet) vertically and 1.5 m (five feet) horizontally on level ground.

If the soil is saturated, or if the rainfall rate is greater than the rate at which water can infiltrate into the soil, surface runoff occurs. If the runoff has sufficient flow energy, it will transport loosened soil particles (sediment) down the slope. *Sheet erosion* is the transport of loosened soil particles by overland flow.

A spoil tip covered in rills and gullies due to erosion processes caused by rainfall: Rummu, Estonia.

Rill erosion refers to the development of small, ephemeral concentrated flow paths which function as both sediment source and sediment delivery systems for erosion on hillslopes. Generally, where water erosion rates on disturbed upland areas are greatest, rills are active. Flow depths in rills are typically of the order of a few centimetres (about an inch) or less and along-channel slopes may be quite steep. This means that rills exhibit hydraulic physics very different from water flowing through the deeper, wider channels of streams and rivers.

Gully erosion occurs when runoff water accumulates and rapidly flows in narrow channels during or immediately after heavy rains or melting snow, removing soil to a considerable depth.

Rivers and Streams

Valley or *stream erosion* occurs with continued water flow along a linear feature. The erosion is both downward, deepening the valley, and headward, extending the valley into the hillside, creating head cuts and steep banks. In the earliest stage of stream erosion, the erosive activity is dominantly vertical, the valleys have a typical V cross-section and the stream gradient is relatively steep. When some base level is reached, the erosive activity switches to lateral erosion, which widens the valley floor and creates a narrow floodplain. The stream gradient becomes nearly flat, and lateral deposition of sediments becomes important as the stream meanders across the valley floor. In all stages of stream erosion, by far the most erosion occurs during times of flood, when more and faster-moving water is available to carry a larger sediment load. In such processes, it is not the water alone that erodes: suspended abrasive particles, pebbles and boulders can also act erosively as they traverse a surface, in a process known as *traction*.

Dobbingstone Burn, Scotland, showing two different types of erosion affecting the same place. Valley erosion is occurring due to the flow of the stream, and the boulders and stones (and much of the soil) that are lying on the stream's banks are glacial till that was left behind as ice age glaciers flowed over the terrain.

Bank erosion is the wearing away of the banks of a stream or river. This is distinguished from changes on the bed of the watercourse, which is referred to as *scour*. Erosion and changes in the form of river banks may be measured by inserting metal rods into the bank and marking the position of the bank surface along the rods at different times.

Thermal erosion is the result of melting and weakening permafrost due to moving water. It can occur both along rivers and at the coast. Rapid river channel migration observed in the Lena River of Siberia is due to thermal erosion, as these portions of the banks are composed of permafrost-cemented non-cohesive materials. Much of this erosion occurs as the weakened banks fail in large slumps. Thermal erosion also affects the Arctic coast, where wave action and near-shore temperatures combine to undercut permafrost bluffs along the shoreline and cause them to fail. Annual erosion rates along a 100-kilometre (62-mile) segment of the Beaufort Sea shoreline averaged 5.6 metres (18 feet) per year from 1955 to 2002.

Coastal Erosion

Wave cut platform caused by erosion of cliffs by the sea, at Southerndown in South Wales.

Erosion of the boulder clay (of Pleistocene age) along cliffs of Filey Bay, Yorkshire, England.

Shoreline erosion, which occurs on both exposed and sheltered coasts, primarily occurs through the action of currents and waves but sea level (tidal) change can also play a role.

Hydraulic action takes place when air in a joint is suddenly compressed by a wave closing the entrance of the joint. This then cracks it. *Wave pounding* is when the sheer energy of the wave hitting the cliff or rock breaks pieces off. *Abrasion* or *corrasion* is caused by waves launching seaload at the cliff. It is the most effective and rapid form of shoreline erosion. *Corrosion* is the dissolving of rock by carbonic acid in sea water. Limestone cliffs are particularly vulnerable to this kind of erosion. *Attrition* is where particles/seaload carried by the waves are worn down as they hit each other and the cliffs. This then makes the material easier to wash away. The material ends up as shingle and sand. Another significant source of erosion, particularly on carbonate coastlines, is the boring, scraping and grinding of organisms, a process termed *bioerosion*.

Sediment is transported along the coast in the direction of the prevailing current (longshore drift). When the upcurrent amount of sediment is less than the amount being carried away, erosion occurs. When the upcurrent amount of sediment is greater, sand or gravel banks will tend to form as a result of deposition. These banks may slowly migrate along the coast in the direction of the longshore drift, alternately protecting and exposing parts of the coastline. Where there is a bend in the coastline, quite often a buildup of eroded material occurs forming a long narrow bank (a spit). Armoured beaches and submerged offshore sandbanks may also protect parts of a coastline from erosion. Over the years, as the shoals gradually shift, the erosion may be redirected to attack different parts of the shore.

Chemical Erosion

Chemical erosion is the loss of matter in a landscape in the form of solutes. Chemical erosion is usually calculated from the solutes found in streams. Anders Rapp pioneered the study of chemical erosion in his work about Kärkevagge published in 1960.

Glaciers

Glacial moraines above Lake Louise, in Alberta, Canada.

Glaciers erode predominantly by three different processes: abrasion/scouring, plucking, and ice thrusting. In an abrasion process, debris in the basal ice scrapes along the bed, polishing and gouging the underlying rocks, similar to sandpaper on wood. Glaciers can also cause pieces of bedrock to crack off in the process of plucking. In ice thrusting, the glacier freezes to its bed, then as it surges forward, it moves large sheets of frozen sediment at the base along with the glacier. This method produced some of the many thousands of lake basins that dot the edge of the Canadian Shield. The erosion caused by glaciers worldwide erodes mountains so effectively that the term *glacial buzz-saw* has become widely used, which describes the limiting effect of glaciers on the height of mountain ranges. As mountains grow higher, they generally allow for more glacial activity (especially in the accumulation zone above the glacial equilibrium line altitude), which causes increased rates of erosion of the mountain, decreasing mass faster than isostatic rebound can add to the mountain. This provides a good example of a negative feedback loop. Ongoing research is showing that while glaciers tend to decrease mountain size, in some areas, glaciers can actually reduce the rate of erosion, acting as a *glacial armour*.

These processes, combined with erosion and transport by the water network beneath the glacier, leave moraines, drumlins, ground moraine (till), kames, kame deltas, moulins, and glacial erratics in their wake, typically at the terminus or during glacier retreat.

Floods

At extremely high flows, kolks, or vortices are formed by large volumes of rapidly rushing water. Kolks cause extreme local erosion, plucking bedrock and creating pothole-type geographical features called Rock-cut basins. Examples can be seen in the flood regions result from glacial Lake Missoula, which created the channeled scablands in the Columbia Basin region of eastern Washington.

Wind Erosion

Wind erosion is a major geomorphological force, especially in arid and semi-arid regions. It is also a major source of land degradation, evaporation, desertification, harmful airborne dust, and crop damage—especially after being increased far above natural rates by human activities such as deforestation, urbanization, and agriculture.

Árbol de Piedra, a rock formation in the Altiplano, Bolivia sculpted by wind erosion.

Wind erosion is of two primary varieties: *deflation*, where the wind picks up and carries away loose particles; and *abrasion*, where surfaces are worn down as they are struck by airborne particles carried by wind. Deflation is divided into three categories: (1) *surface creep*, where larger, heavier particles slide or roll along the ground; (2) *saltation*, where particles are lifted a short height into the air, and bounce and saltate across the surface of the soil; and (3) *suspension*, where very small and light particles are lifted into the air by the wind, and are often carried for long distances. Saltation is responsible for the majority (50-70%) of wind erosion, followed by suspension (30-40%), and then surface creep (5-25%).

Wind erosion is much more severe in arid areas and during times of drought. For example, in the Great Plains, it is estimated that soil loss due to wind erosion can be as much as 6100 times greater in drought years than in wet years.

Mass Movement

Wadi in Makhtesh Ramon, Israel, showing gravity collapse erosion on its banks.

Mass movement is the downward and outward movement of rock and sediments on a sloped surface, mainly due to the force of gravity.

Mass movement is an important part of the erosional process, and is often the first stage in the breakdown and transport of weathered materials in mountainous areas. It moves material from higher elevations to lower elevations where other eroding agents such as streams and glaciers can then pick up the material and move it to even lower elevations. Mass-movement processes are always occurring continuously on all slopes; some mass-movement processes act very slowly; others occur very suddenly, often with disastrous results. Any perceptible down-slope movement of rock or sediment is often referred to in general terms as a landslide. However, landslides can be classified in a much more detailed way that reflects the mechanisms responsible for the movement and the velocity at which the movement occurs. One of the visible topographical manifestations of a very slow form of such activity is a scree slope.

Slumping happens on steep hillsides, occurring along distinct fracture zones, often within materials like clay that, once released, may move quite rapidly downhill. They will often show a spoon-shaped isostatic depression, in which the material has begun to slide downhill. In some cases, the slump is caused by water beneath the slope weakening it. In many cases it is simply the result of poor engineering along highways where it is a regular occurrence. *Surface creep* is the slow movement of soil and rock debris by gravity which is usually not perceptible except through extended observation. However, the term can also describe the rolling of dislodged soil particles 0.5 to 1.0 mm (0.02 to 0.04 in) in diameter by wind along the soil surface.

Factors Affecting Erosion Rates

Climate

The amount and intensity of precipitation is the main climatic factor governing soil erosion by water. The relationship is particularly strong if heavy rainfall occurs at times when, or in locations where, the soil's surface is not well protected by vegetation. This might be during periods when agricultural activities leave the soil bare, or in semi-arid regions where vegetation is naturally sparse. Wind erosion requires strong winds, particularly during times of drought when vegetation is sparse and soil is dry (and so is more erodible). Other climatic factors such as average temperature and temperature range may also affect erosion, via their effects on vegetation and soil properties. In general, given similar vegetation and ecosystems, areas with more precipitation (especially high-intensity rainfall), more wind, or more storms are expected to have more erosion.

In some areas of the world (e.g. the mid-western USA), rainfall intensity is the primary determinant of erosivity, with higher intensity rainfall generally resulting in more soil erosion by water. The size and velocity of rain drops is also an important factor. Larger and higher-velocity rain drops have greater kinetic energy, and thus their impact will displace soil particles by larger distances than smaller, slower-moving rain drops.

In other regions of the world (e.g. western Europe), runoff and erosion result from relatively low intensities of stratiform rainfall falling onto previously saturated soil. In such situations, rainfall amount rather than intensity is the main factor determining the severity of soil erosion by water.

Vegetative Cover

Vegetation acts as an interface between the atmosphere and the soil. It increases the permeability of the soil to rainwater, thus decreasing runoff. It shelters the soil from winds, which results in decreased wind erosion, as well as advantageous changes in microclimate. The roots of the plants bind the soil together, and interweave with other roots, forming a more solid mass that is less susceptible to both water and wind erosion. The removal of vegetation increases the rate of surface erosion.

Topography

The topography of the land determines the velocity at which surface runoff will flow, which in turn determines the erosivity of the runoff. Longer, steeper slopes (especially those without adequate vegetative cover) are more susceptible to very high rates of erosion during heavy rains than shorter, less steep slopes. Steeper terrain is also more prone to mudslides, landslides, and other forms of gravitational erosion processes.

Tectonics

Tectonic processes control rates and distributions of erosion at the Earth's surface. If tectonic action causes part of the Earth's surface (e.g., a mountain range) to be raised or lowered relative to surrounding areas, this must necessarily change the gradient of the land surface. Because erosion rates are almost always sensitive to local slope, this will change the rates of erosion in the uplifted area. Active tectonics also brings fresh, unweathered rock towards the surface, where it is exposed to the action of erosion.

However, erosion can also affect tectonic processes. The removal by erosion of large amounts of rock from a particular region, and its deposition elsewhere, can result in a lightening of the load on the lower crust and mantle. Because tectonic processes are driven by gradients in the stress field developed in the crust, this unloading can in turn cause tectonic or isostatic uplift in the region. In some cases, it has been hypothesised that these twin feedbacks can act to localise and enhance zones of very rapid exhumation of deep crustal rocks beneath places on the Earth's surface with extremely high erosion rates, for example, beneath the extremely steep terrain of Nanga Parbat in the western Himalayas. Such a place has been called a "tectonic aneurysm".

Erosion of Earth Systems

Mountain Ranges

Mountain ranges are known to take many millions of years to erode to the degree they effectively cease to exist. Scholars Pitman and Golovchenko estimate that it takes probably more than 450 million years to erode a mountain mass similar to the Himalaya into an almost-flat peneplain if there are no major sea-level changes. Erosion of mountains massifs can create a pattern of equally high summits called summit accordance. It has been argued that extension during post-orogenic collapse is a more effective mechanism of lowering the height of orogenic mountains than erosion.

Examples of heavily eroded mountain ranges include the Timanides of Northern Russia. Erosion of this orogen has produced sediments that are now found in the East European Platform, including the Cambrian Sablya Formation near Lake Ladoga. Studies of these sediments points that its likely that the erosion of the orogen was beginning in the Cambrian and then became stronger in Ordovician.

Soils

If the rate of erosion is higher than the rate of soil formation the soils are being destroyed by erosion. Where soil is not destroyed by erosion, erosion can in some cases prevent the formation of soil features that form slowly. Inceptisols are common soils that form in areas of fast erosion.

While erosion of soils is a natural process, human activities have increased by 10-40 times the rate at which erosion is occurring globally. Excessive (or accelerated) erosion causes both "on-site" and "off-site" problems. On-site impacts include decreases in agricultural productivity and (on natural landscapes) ecological collapse, both because of loss of the nutrient-rich upper soil layers. In some cases, the eventual end result is desertification. Off-site effects include sedimentation of waterways and eutrophication of water bodies, as well as sediment-related damage to roads and houses. Water and wind erosion are the two primary causes of land degradation; combined, they are responsible for about 84% of the global extent of degraded land, making excessive erosion one of the most significant environmental problems worldwide.

Diagenesis

Diagenesis is the change of sediments or existing sedimentary rocks into a different sedimentary rock during and after rock formation (lithification), at temperatures and pressures less than that required for the formation of metamorphic rocks. It does not include changes from weathering. It is any chemical, physical, or biological change undergone by a sediment after its initial deposition, after its lithification. This process excludes surface alteration (weathering) and metamorphism. These changes happen at relatively low temperatures and pressures and result in changes to the rock's original mineralogy and texture. There is no sharp boundary between diagenesis and metamorphism, but the latter occurs at higher temperatures and pressures than the former.

Hydrothermal solutions, meteoric groundwater, porosity, permeability, solubility, and time are all influential factors.

After deposition, sediments are compacted as they are buried beneath successive layers of sediment and cemented by minerals that precipitate from solution. Grains of sediment, rock fragments and fossils can be replaced by other minerals during diagenesis. Porosity usually decreases during diagenesis, except in rare cases such as dissolution of minerals and dolomitization.

The study of diagenesis in rocks is used to understand the geologic history they have undergone and the nature and type of fluids that have circulated through them. From a commercial standpoint, such studies aid in assessing the likelihood of finding various economically viable mineral and hydrocarbon deposits.

The process of diagenesis is also important in the decomposition of bone tissue.

The Role of Diagenesis in Anthropology and Paleontology

Originally calcitic crinoid stem (in cross-section) diagenetically replaced by marcasite in a siderite concretion; Lower Carboniferous.

The term diagenesis, literally meaning "across generation", is extensively used in geology. However, this term has filtered into the field of anthropology, archaeology and paleontology to describe the changes and alterations that take place on skeletal (biological) material. Specifically, diagenesis "is the cumulative physical, chemical and biological environment; these processes will modify an organic object's original chemical and/or structural properties and will govern its ultimate fate, in terms of preservation or destruction". In order to assess the potential impact of diagenesis on archaeological or fossil bones, many factors need to be assessed, beginning with elemental and mineralogical composition of bone and enveloping soil, as well as the local burial environment (geology, climatology, groundwater).

The composite nature of bone, comprising one-third organic (mainly protein collagen) and two thirds mineral (calcium phosphate mostly in the form of hydroxyapatite) renders its diagenesis more complex. Alteration occurs at all scales from molecular loss and substitution, through crystallite reorganization, porosity and microstructural changes, and in many cases, to disintegration of the complete unit. Three general pathways of the diagenesis of bone have been identified:

1. chemical deterioration of the organic phase.
2. chemical deterioration of the mineral phase.
3. (micro) biological attack of the composite.

They are as follows:

1. The dissolution of collagen depends on time, temperature and environmental pH. At high temperatures, the rate of collagen loss will be accelerated and extreme pH can cause collagen swelling and accelerated hydrolysis. Due to the increase in porosity of bones through collagen loss, the bone becomes susceptible to hydrolytic infiltration where the hydroxyapatite, with its affinity for amino acids, permits charged species of endogenous and exogenous origin to take up residence.
2. The hydrolytic activity plays a key role in the mineral phase transformations that exposes the collagen to accelerated chemical- and bio-degradation. Chemical changes affect crystallinity. Mechanisms of chemical change, such as the uptake of F^- or CO_3^- may cause recrystallization where hydroxyapatite is dissolved and re-precipitated allowing for the incorporation of substitution of exogenous material.
3. Once an individual has been interred, microbial attack, the most common mechanism of bone deterioration, occurs rapidly. During this phase, most bone collagen is lost and porosity is increased. The dissolution of the mineral phase caused by low pH permits access to the collagen by extracellular microbial enzymes thus microbial attack.

The Role of Diagenesis in Hydrocarbon Generation

When animal or plant matter is buried during sedimentation, the constituent organic molecules (lipids, proteins, carbohydrates and lignin-humic compounds) break down due to the increase in temperature and pressure. This transformation occurs in the first few hundred meters of burial and results in the creation of two primary products: kerogens and bitumens.

It is generally accepted that hydrocarbons are formed by the thermal alteration of these kerogens (the *biogenic* theory). In this way, given certain conditions (which are largely temperature-dependent) kerogens will break down to form hydrocarbons through a chemical process known as cracking, or catagenesis.

A kinetic model based on experimental data can capture most of the essential transformation in diagenesis, and a mathematical model in a compacting porous medium to model the dissolution-precipitation mechanism. These models have been intensively studied and applied in real geological applications.

Diagenesis has been divided, based on hydrocarbon and coal genesis into: *eodiagenesis* (early), *mesodiagenesis* (middle) and *telodiagenesis* (late). During the early or eodiagenesis stage shales lose pore water, little to no hydrocarbons are formed and coal varies between lignite and sub-bituminous. During mesodiagenesis, dehydration of clay minerals occurs, the main development of oil genesis occurs and high to low volatile bituminous coals are formed. During telodiagenesis organic matter undergoes cracking and dry gas is produced; semi-anthracite coals develop.

Early diagenesis in newly formed aquatic sediments is mediated by microorganisms using different electron acceptors as part of their metabolism. Organic matter is mineralized, liberating gaseous carbon dioxide (CO_2) in the porewater, which, depending on the conditions, can diffuse into the water column. The various processes of mineralization in this phase are nitrification and denitrification, manganese oxide reduction, iron hydroxide reduction, sulfate reduction, and fermentation.

The Role of Diagenesis in Bone Decomposition

Diagenesis alters the proportions of organic collagen and inorganic components (hydroxyapatite, calcium, magnesium) of bone exposed to environmental conditions, especially moisture. This is accomplished by the exchange of natural bone constituents, deposition in voids or defects, adsorption onto the bone surface and leaching from the bone.

Weathering

Weathering is the breaking down of rocks, soil, and minerals as well as wood and artificial materials through contact with the Earth's atmosphere, waters, and biological organisms. Weathering occurs *in situ* (on site), that is, in the same place, with little or no movement, and thus should not be confused with erosion, which involves the movement of rocks and minerals by agents such as water, ice, snow, wind, waves and gravity and then being transported and deposited in other locations.

Two important classifications of weathering processes exist – physical and chemical weathering; each sometimes involves a biological component. Mechanical or physical weathering involves the breakdown of rocks and soils through direct contact with atmospheric conditions, such as heat, water, ice and pressure. The second classification, chemical weathering, involves the direct effect of atmospheric chemicals or biologically produced chemicals also known as biological weathering in the breakdown of rocks, soils and minerals. While physical weathering is accentuated in very cold or very dry environments, chemical reactions are most intense where the climate is wet and hot. However, both types of weathering occur together, and each tends to accelerate the other. For example, physical abrasion (rubbing together) decreases the size of particles and therefore increases their surface area, making them more susceptible to rapid chemical reactions. The various agents act in concert to convert primary minerals (feldspars and micas) to secondary minerals (clays and carbonates) and release plant nutrient elements in soluble forms.

The materials left over after the rock breaks down combined with organic material creates soil. The mineral content of the soil is determined by the parent material; thus, a soil derived from a single rock type can often be deficient in one or more minerals needed for good fertility, while a soil weathered from a mix of rock types (as in glacial, aeolian or alluvial sediments) often makes more fertile soil. In addition, many of Earth's landforms and landscapes are the result of weathering processes combined with erosion and re-deposition.

Physical Weathering

Physical weathering, also recognized as mechanical weathering, is the class of processes that causes the disintegration of rocks without chemical change. The primary process in physical weathering is abrasion (the process by which clasts and other particles are reduced in size). However, chemical and physical weathering often go hand in hand. Physical weathering can occur due to temperature, pressure, frost etc. For example, cracks exploited by physical weathering will increase the surface area exposed to chemical action, thus amplifying the rate of disintegration.

Abrasion by water, ice, and wind processes loaded with sediment can have tremendous cutting power, as is amply demonstrated by the gorges, ravines, and valleys around the world. In glacial areas, huge moving ice masses embedded with soil and rock fragments grind down rocks in their path and carry away large volumes of material. Plant roots sometimes enter cracks in rocks and pry them apart, resulting in some disintegration; Burrowing animals may help disintegrate rock through their physical action. However, such influences are usually of little importance in producing parent material when compared to the drastic physical effects of water, ice, wind, and temperature change. Physical weathering is also called mechanical weathering or disaggregation.

Thermal Stress

Thermal stress weathering (sometimes called insolation weathering) results from the expansion and contraction of rock, caused by temperature changes. For example, heating of rocks by sunlight or fires can cause expansion of their constituent minerals. As some minerals expand more than others, temperature changes set up differential stresses that eventually cause the rock to crack apart. Because the outer surface of a rock is often warmer or colder than the more protected inner portions, some rocks may weather by exfoliation – the peeling away of outer layers. This process may be sharply accelerated if ice forms in the surface cracks. When water freezes, it expands with a force of about 1465 Mg/m^2, disintegrating huge rock masses and dislodging mineral grains from smaller fragments.

Thermal stress weathering comprises two main types, thermal shock and thermal fatigue. Thermal stress weathering is an important mechanism in deserts, where there is a large diurnal temperature range, hot in the day and cold at night. The repeated heating and cooling exerts stress on the outer layers of rocks, which can cause their outer layers to peel off in thin sheets. The process of peeling off is also called exfoliation. Although temperature changes are the principal driver, moisture can enhance thermal expansion in rock. Forest fires and range fires are also known to cause significant weathering of rocks and boulders exposed along the ground surface. Intense localized heat can rapidly expand a boulder.

The thermal heat from wildfire can cause significant weathering of rocks and boulders, heat can rapidly expand a boulder and thermal shock can occur. The differential expansion of a thermal gradient can be understood in terms of stress or of strain, equivalently. At some point, this stress can exceed the strength of the material, causing a crack to form. If nothing stops this crack from propagating through the material, it will result in the object's structure to fail.

Frost Weathering

A rock in Abisko, Sweden fractured along existing joints possibly by frost weathering or thermal stress.

Frost weathering, frost wedging, *ice wedging* or *cryofracturing* is the collective name for several processes where ice is present. These processes include frost shattering, frost-wedging and freeze-thaw weathering. Severe frost shattering produces huge piles of rock fragments called scree which may be located at the foot of mountain areas or along slopes. Frost weathering is common in mountain areas where the temperature is around the freezing point of water. Certain frost-susceptible soils expand or heave upon freezing as a result of water migrating via capillary action to grow ice lenses near the freezing front. This same phenomenon occurs within pore spaces of rocks. The ice accumulations grow larger as they attract liquid water from the surrounding pores. The ice crystal growth weakens the rocks which, in time, break up. It is caused by the approximately 10% (9.87) expansion of ice when water freezes, which can place considerable stress on anything containing the water as it freezes.

Freeze induced weathering action occurs mainly in environments where there is a lot of moisture, and temperatures frequently fluctuate above and below freezing point, especially in alpine and periglacial areas. An example of rocks susceptible to frost action is chalk, which has many pore spaces for the growth of ice crystals. This process can be seen in Dartmoor where it results in the formation of tors. When water that has entered the joints freezes, the ice formed strains the walls of the joints and causes the joints to deepen and widen. When the ice thaws, water can flow further into the rock. Repeated freeze-thaw cycles weaken the rocks which, over time, break up along the joints into angular pieces. The angular rock fragments gather at the foot of the slope to form a talus slope (or scree slope). The splitting of rocks along the joints into blocks is called block disintegration. The blocks of rocks that are detached are of various shapes depending on rock structure.

Ocean Waves

Wave action and water chemistry lead to structural failure in exposed rocks.

Coastal geography is formed by the weathering of wave actions over geological times or can happen more abruptly through the process of salt weathering.

Pressure Release

In pressure release, also known as unloading, overlying materials (not necessarily rocks) are removed (by erosion, or other processes), which causes underlying rocks to expand and fracture parallel to the surface.

Pressure release could have caused the exfoliated granite sheets shown in the picture.

Intrusive igneous rocks (e.g. granite) are formed deep beneath the Earth's surface. They are under tremendous pressure because of the overlying rock material. When erosion removes the overlying rock material, these intrusive rocks are exposed and the pressure on them is released. The outer parts of the rocks then tend to expand. The expansion sets up stresses which cause fractures parallel to the rock surface to form. Over time, sheets of rock break away from the exposed rocks along the fractures, a process known as exfoliation. Exfoliation due to pressure release is also known as "sheeting".

Retreat of an overlying glacier can also lead to exfoliation due to pressure release.

Salt-crystal Growth

Tafoni at Salt Point State Park, Sonoma County, California.

Salt crystallization, otherwise known as haloclasty, causes disintegration of rocks when saline solutions seep into cracks and joints in the rocks and evaporate, leaving salt crystals behind. These salt crystals expand as they are heated up, exerting pressure on the confining rock.

Salt crystallization may also take place when solutions decompose rocks (for example, limestone and chalk) to form salt solutions of sodium sulfate or sodium carbonate, of which the moisture evaporates to form their respective salt crystals.

The salts which have proved most effective in disintegrating rocks are sodium sulfate, magnesium sulfate, and calcium chloride. Some of these salts can expand up to three times or even more.

It is normally associated with arid climates where strong heating causes strong evaporation and therefore salt crystallization. It is also common along coasts. An example of salt weathering can be seen in the honeycombed stones in sea wall. Honeycomb is a type of tafoni, a class of cavernous rock weathering structures, which likely develop in large part by chemical and physical salt weathering processes.

Biological Effects on Mechanical Weathering

Living organisms may contribute to mechanical weathering (as well as chemical weathering, see 'biological' weathering below). Lichens and mosses grow on essentially bare rock surfaces and create a more humid chemical microenvironment. The attachment of these organisms to the rock surface enhances physical as well as chemical breakdown of the surface microlayer of the rock. On a larger scale, seedlings sprouting in a crevice and plant roots exert physical pressure as well as providing a pathway for water and chemical infiltration.

Chemical Weathering

Comparison of unweathered (left) and weathered (right) limestone.

Chemical weathering changes the composition of rocks, often transforming them when water interacts with minerals to create various chemical reactions. Chemical weathering is a gradual and ongoing process as the mineralogy of the rock adjusts to the near surface environment. New or *secondary minerals* develop from the original minerals of the rock. In this the processes of oxidation and hydrolysis are most important. Chemical weathering is enhanced by such geological agents as the presence of water and oxygen, as well as by such biological agents as the acids produced by microbial and plant-root metabolism.

The process of mountain block uplift is important in exposing new rock strata to the atmosphere and moisture, enabling important chemical weathering to occur; significant release occurs of Ca^{2+} and other ions into surface waters.

Dissolution and Carbonation

Rainfall is acidic because atmospheric carbon dioxide dissolves in the rainwater producing weak carbonic acid. In unpolluted environments, the rainfall pH is around 5.6.

Acid rain occurs when gases such as sulfur dioxide and nitrogen oxides are present in the atmosphere. These oxides react in the rain water to produce stronger acids and can lower the pH to 4.5 or even 3.0. Sulfur dioxide, SO_2, comes from volcanic eruptions or from fossil fuels, can become sulfuric acid within rainwater, which can cause solution weathering to the rocks on which it falls.

A pyrite cube has dissolved away from host rock, leaving gold behind.

Some minerals, due to their natural solubility (e.g. evaporites), oxidation potential (iron-rich minerals, such as pyrite), or instability relative to surficial conditions will weather through dissolution naturally, even without acidic water.

One of the most well-known solution weathering processes is carbonation, the process in which atmospheric carbon dioxide leads to solution weathering. Carbonation occurs on rocks which contain calcium carbonate, such as limestone and chalk. This takes place when rain combines with carbon dioxide or an organic acid to form a weak carbonic acid which reacts with calcium carbonate (the limestone) and forms calcium bicarbonate. This process speeds up with a decrease in temperature, not because low temperatures generally drive reactions faster, but because colder water holds more dissolved carbon dioxide gas. Carbonation is therefore a large feature of glacial weathering.

The reactions as follows:

$$CO_2 + H_2O \rightarrow H_2CO_3$$

carbon dioxide + water → carbonic acid

$$H_2CO_3 + CaCO_3 \rightarrow Ca(HCO_3)_2$$

carbonic acid + calcium carbonate → calcium bicarbonate

Carbonation on the surface of well-jointed limestone produces a dissected limestone pavement. This process is most effective along the joints, widening and deepening them.

Hydration

Olivine weathering to iddingsite within a mantle xenolith.

Mineral hydration is a form of chemical weathering that involves the rigid attachment of H+ and OH- ions to the atoms and molecules of a mineral.

When rock minerals take up water, the increased volume creates physical stresses within the rock. For example, iron oxides are converted to iron hydroxides and the hydration of anhydrite forms gypsum.

A freshly broken rock shows differential chemical weathering (probably mostly oxidation) progressing inward. This piece of sandstone was found in glacial drift near Angelica, New York.

Hydrolysis on Silicates and Carbonates

Hydrolysis is a chemical weathering process affecting silicate and carbonate minerals. In such reactions, pure water ionizes slightly and reacts with silicate minerals. An example reaction:

$$Mg_2SiO_4 + 4\,H^+ + 4\,OH^- \rightleftharpoons 2\,Mg^{2+} + 4\,OH^- + H_4SiO_4$$

olivine (forsterite) + four ionized water molecules ⇌ ions in solution + silicic acid in solution

This reaction theoretically results in complete dissolution of the original mineral, if enough water is available to drive the reaction. In reality, pure water rarely acts as a H^+ donor. Carbon dioxide, though, dissolves readily in water forming a weak acid and H^+ donor.

$$Mg_2SiO_4 + 4\ CO_2 + 4\ H_2O \rightleftharpoons 2\ Mg^{2+} + 4\ HCO_3^- + H_4SiO_4$$

olivine (forsterite) + carbon dioxide + water ⇌ Magnesium and bicarbonate ions in solution + silicic acid in solution

This hydrolysis reaction is much more common. Carbonic acid is consumed by silicate weathering, resulting in more alkaline solutions because of the bicarbonate. This is an important reaction in controlling the amount of CO_2 in the atmosphere and can affect climate.

Aluminosilicates when subjected to the hydrolysis reaction produce a secondary mineral rather than simply releasing cations.

$$2\ KAlSi_3O_8 + 2\ H_2CO_3 + 9\ H_2O \rightleftharpoons Al_2Si_2O_5(OH)_4 + 4\ H_4SiO_4 + 2\ K^+ + 2\ HCO_3^-$$

Orthoclase (aluminosilicate feldspar) + carbonic acid + water ⇌ Kaolinite (a clay mineral) + silicic acid in solution + potassium and bicarbonate ions in solution

Oxidation

Oxidized pyrite cubes.

Within the weathering environment chemical oxidation of a variety of metals occurs. The most commonly observed is the oxidation of Fe^{2+} (iron) and combination with oxygen and water to form Fe^{3+} hydroxides and oxides such as goethite, limonite, and hematite. This gives the affected rocks a reddish-brown coloration on the surface which crumbles easily and weakens the rock. This process is better known as 'rusting', though it is distinct from the rusting of metallic iron. Many other metallic ores and minerals oxidize and hydrate to produce colored deposits, such as chalcopyrites or $CuFeS_2$ oxidizing to copper hydroxide and iron oxides.

Biological Weathering

A number of plants and animals may create chemical weathering through release of acidic compounds, i.e. the effect of moss growing on roofs is classed as weathering.

Mineral weathering can also be initiated and/or accelerated by soil microorganisms. Lichens on rocks are thought to increase chemical weathering rates. For example, an experimental study on hornblende granite in New Jersey, USA, demonstrated a 3x – 4x increase in weathering rate under lichen covered surfaces compared to recently exposed bare rock surfaces.

Biological weathering of basalt by lichen, La Palma.

The most common forms of biological weathering are the release of chelating compounds (i.e. organic acids, siderophores) and of acidifying molecules (i.e. protons, organic acids) by plants so as to break down aluminium and iron containing compounds in the soils beneath them. Decaying remains of dead plants in soil may form organic acids which, when dissolved in water, cause chemical weathering. Extreme release of chelating compounds can easily affect surrounding rocks and soils, and may lead to podsolisation of soils.

The symbiotic mycorrhizal fungi associated with tree root systems can release inorganic nutrients from minerals such as apatite or biotite and transfer these nutrients to the trees, thus contributing to tree nutrition. It was also recently evidenced that bacterial communities can impact mineral stability leading to the release of inorganic nutrients. To date a large range of bacterial strains or communities from diverse genera have been reported to be able to colonize mineral surfaces and/or to weather minerals, and for some of them a plant growth promoting effect was demonstrated. The demonstrated or hypothesised mechanisms used by bacteria to weather minerals include several oxidoreduction and dissolution reactions as well as the production of weathering agents, such as protons, organic acids and chelating molecules.

Building Weathering

Buildings made of any stone, brick or concrete are susceptible to the same weathering agents as any exposed rock surface. Also statues, monuments and ornamental stonework can be badly damaged by natural weathering processes. This is accelerated in areas severely affected by acid rain.

Properties of Well-weathered Soils

Three groups of minerals often remain in well-weathered soils: silicate clays, very resistant end products including iron and aluminium oxide clays, and very resistant primary minerals such as quartz. In highly weathered soils of humid tropical and subtropical regions, the oxides of iron and aluminium, and certain silicate clays with low Si/Al ratios, predominate because most other constituents have been broken down and removed.

References

- Nancy D. Gordon (2004-06-01). “Erosion and Scour”. Stream hydrology: an introduction for ecologists. ISBN 978-0-470-84357-4.
- Balba, A. Monem “Desertification: Wind erosion”. Management of Problem Soils in Arid Ecosystems. CRC Press. p. 214. ISBN 978-0-87371-811-0.
- Blanco-Canqui, Humberto; Rattan, Lal (2008). “Wind erosion”. Principles of soil conservation and management. Dordrecht: Springer. pp. 54–80. ISBN 9781402087097.
- Blanco, Humberto & Lal, Rattan (2010). “Soil and water conservation”. Principles of Soil Conservation and Management. Springer. p. 2. ISBN 978-90-481-8529-0.
- Wiggs, Giles F.S. (2011). “Geomorphological hazards in drylands”. In Thomas, David S.G. Arid Zone Geomorphology: Process, Form and Change in Drylands. John Wiley & Sons. p. 588. ISBN 978-0-470-71076-0.
- Blanco-Canqui, Humberto; Rattan, Lal (2008). “Soil and water conservation”. Principles of soil conservation and management. Dordrecht: Springer. pp. 1–20. ISBN 9781402087097.
- Blanco-Canqui, Humberto; Rattan, Lal (2008). “Water erosion”. Principles of soil conservation and management. Dordrecht: Springer. pp. 21–53. ISBN 9781402087097.
- Van Beek, Rens (2008). Hillside processes: mass wasting, slope stability, and erosion . In Norris, Joanne E. et al. Slope Stability and Erosion Control: Ecotechnological Solutions. Springer. ISBN 978-1-4020-6675-7
- Boardman, John; Poesen, Jean, eds. (2007). Soil Erosion in Europe. Chichester: John Wiley & Sons. ISBN 9780470859117.
- Toy, Terrence J.; Foster, George R.; Renard, Kenneth G. (2002). Soil erosion : processes, prediction, measurement, and control. New York, NY: Wiley. ISBN 9780471383697.
- Borah, Deva K. et al. (2008). “Watershed sediment yield”. In Garcia, Marcelo H. Sedimentation Engineering: Processes, Measurements, Modeling, and Practice. ASCE Publishing. p. 828. ISBN 978-0-7844-0814-8.
- Cornelis, Wim S. (2006). “Hydroclimatology of wind erosion in arid and semi-arid” environments”. In D’Odorico, Paolo & Porporato, Amilcare. Dryland Ecohydrology. Springer. p. 141. ISBN 978-1-4020-4261-4.
- Gray, Donald H. & Sotir, Robbin B. “Surficial erosion and mass movement”. Biotechnical and Soil Bioengineering Slope Stabilization: A Practical Guide for Erosion Control. John Wiley & Sons. p. 20. ISBN 978-0-471-04978-4.
- Zheng, Xiaojing & Huang, Ning (2009). Mechanics of Wind-Blown Sand Movements. Springer. pp. 7–8. ISBN 978-3-540-88253-4.

- Wainwright, John & Brazier, Richard E. (2011). "Slope systems". In Thomas, David S.G. Arid Zone Geomorphology: Process, Form and Change in Drylands. John Wiley & Sons. ISBN 978-0-470-71076-0.
- Burbank, Douglas W.; Anderson, Robert S. (2011). "Tectonic and surface uplift rates". Tectonic Geomorphology. John Wiley & Sons. pp. 270–271. ISBN 978-1-4443-4504-9.
- Whisenant, Steve G. (2008). "Terrestrial systems". In Perrow Michael R. & Davy, Anthony J. Handbook of Ecological Restoration: Principles of Restoration. Cambridge University Press. p. 89. ISBN 978-0-521-04983-2.

Chapter 4

Sedimentary Structures

The features of sediments and sedimentary rocks, which are formed at the time of deposition, are called sedimentary structures. They can be broadly categorized into primary sedimentary structures and secondary sedimentary structures. This chapter has been carefully written to provide an easy understanding of these types of sedimentary structures.

Sedimentary structures are the larger, generally three-dimensional physical features of sedimentary rocks; they are best seen in outcrop or in large hand specimens rather than through a microscope. Sedimentary structures include features like bedding, ripple marks, fossil tracks and trails, and mud cracks. They conventionally are subdivided into categories based on mode of genesis. Structures that are produced at the same time as the sedimentary rock in which they occur are called primary sedimentary structures. Examples include bedding or stratification, graded bedding, and cross-bedding. Sedimentary structures that are produced shortly after deposition and as a result of compaction and desiccation are called penecontemporaneous sedimentary structures. Examples include mud cracks and load casts. Still other sedimentary structures like concretions, vein fillings, and stylolites form well after deposition and penecontemporaneous modification; these are known as secondary structures. Finally, others like stromatolites and organic burrows and tracks, though they may in fact be primary, penecontemporaneous, or even secondary, may be grouped as a fourth category—organic sedimentary structures.

Considerable attention is paid to the sedimentary structures exhibited by any sedimentary rock. Primary sedimentary structures are particularly useful because their abundance and size suggest the probable transporting and depositional agents. Certain varieties of primary sedimentary structures like cross-bedding and ripple marks display orientations that are consistently related to the direction of current movement. Such structures are referred to as directional sedimentary structures because they can be used to infer the ancient paleocurrent pattern or dispersal system by which a sedimentary rock unit was deposited. Other sedimentary structures are stratigraphic "top and bottom" indicators. For example, the progressive upward decrease in clastic grain size diameters, known as graded bedding, would allow a geologist to determine which way is stratigraphically "up"—i.e., toward the younger beds in a dipping sedimentary bed. The suite (repeated sequence) of sedimentary structures in any single stratigraphic unit is another attribute by which that unit may be physically differentiated from others in the region.

Primary Sedimentary Structures

Predepositional (Interbed) Structures

Predepositional sedimentary structures occur on surfaces between beds. They were formed before the deposition of the overlying bed. The majority of this group of structures are erosional in origin. Before describing these structures, note two points. The first is one of terminology. When the interface between two beds is split open, the convex structures that depend from the upper bed are termed "casts." The concave hollows in the underlying bed into which these fit are termed "molds". The second point to note is that the ease and frequency with which these bottom structures are seen is related to the degree of lithification.

Unconsolidated sediments seldom split along interbed boundaries because of their friable nature. This may explain the apparent scarcity of bottom structures recorded from modern deep-sea turbidite sands. On the other extreme, well-lithified metasediments in fold belts tend to split more readily along subvertical tectonic fractures rather than along bedding planes. Thus bottom structures are most commonly seen in moderately well-lithified sediments that split along bedding surfaces.

Channels

The largest predepositional interbed structures are channels. These may be kilometers wide and hundreds of meters deep. They occur in diverse environments ranging from subaerial alluvial plains to submarine continental margins. Channeling is initiated by localized linear erosion by fluid flow aided by corrosive bed load. Once a channel is established, however, a horizontal component of erosion develops due to undercutting of the channel bank followed by collapse of the overhanging sector.

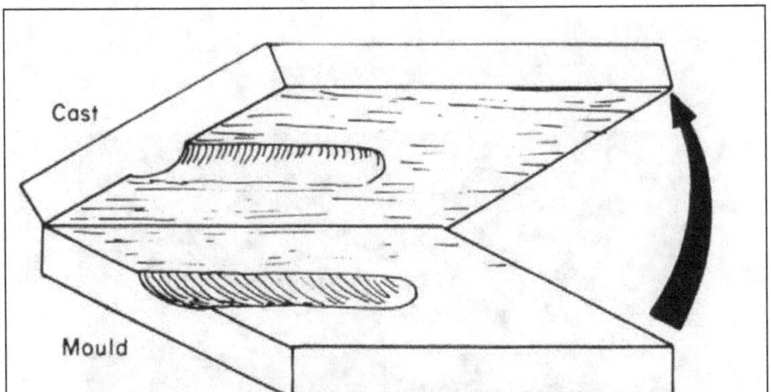

Nomenclature for the occurrence of bed interface sedimentary structures (sole markings).

The best studied channels are those of fluvial systems and particular attention has been paid to the genesis of channel meandering and the mathematical relationships between sinuosity, channel width, depth, gradient, and discharge. In ancient channels, depth, width, sediment grade, and flow direction can easily be established. Sinuosity can also be measured where there is adequate exposure or well control. Using these data, attempts have been made to calculate the gradient and stream power of ancient channel systems.

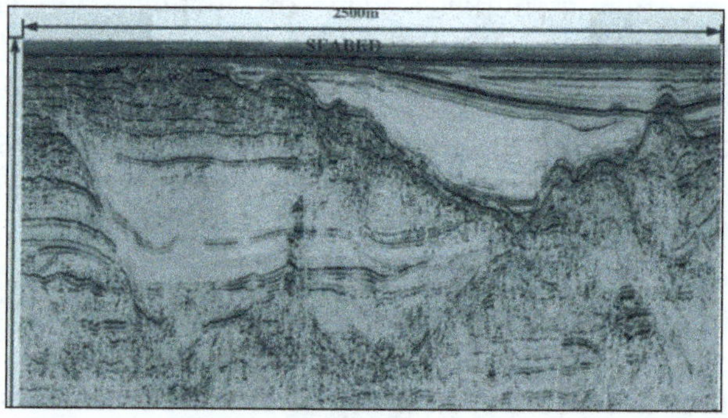

Ultra-high-resolution seismic line showing multistory channel structures. The vertical axis is twoway travel time. The floor of the lower channel is some 100 m below present sea floor.

Channels are of great economic importance for several reasons. They can be petroleum reservoirs and aquifers, they can contain placer and replacement mineral ore bodies, and they can cut out coal seams. Instances are cited throughout this text. Smaller and less dramatic are the interbed structures termed scour-and-fill. These are smallscale channels whose dimensions are measured in decimeters rather than meters. They too occur in diverse environments.

There is a vast nomenclature for the numerous small interbed erosional structures. Reference should be made to the atlases of structures previously cited for exhaustive details of these. The following account describes the three most common varieties: flutes, grooves, and tool marks. Flutes and grooves are scoured by the current alone, aided by granular bed load, whereas tool marks are made by single particles generally of pebble grade.

Flute Marks

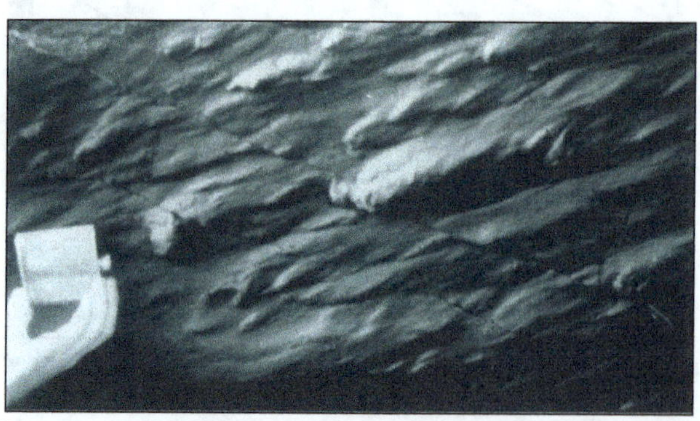

Photograph of flute marks on the base of a turbidite sand. Current flowed from bottom left to top right. Aberystwyth Grits (Lower Silurian), Aberystwyth, Wales.

Flutes are heel-shaped hollows, scoured into mud bottoms. Each hollow is generally infilled by sand, contiguous with the overlying bed. The rounded part of the flute is at the upcurrent end. The flared end points are downcurrent. Flutes are about 1-5 cm wide and 5-20 cm long. They are typically gregarious. Fluting has long been attributed to the localized scouring action of a current moving over an unconsolidated mud bottom. As the current velocity declines, flute erosion ceases and the hollows are buried beneath a bed of sand.

Researchers have has described experiments that explain the hydraulic conditions that generate flutes. The technique used in this and other experiments described in this chapter is as follows. A bed form is carved onto a plaster of Paris (gypsum) surface. Small, regularly spaced pits are marked on this. The slab is then placed in a flume and water passed over it. The pits become elongated in a downcurrent direction. The pit trends show how the current flow, immediately at the bed form:water interface, diverges from the mean flow direction. Allen's experiments show that the flow pattern for flute erosion consists of two horizontal corkscrew vortices that lie beneath a zone of fluid separation at the top of the flute.

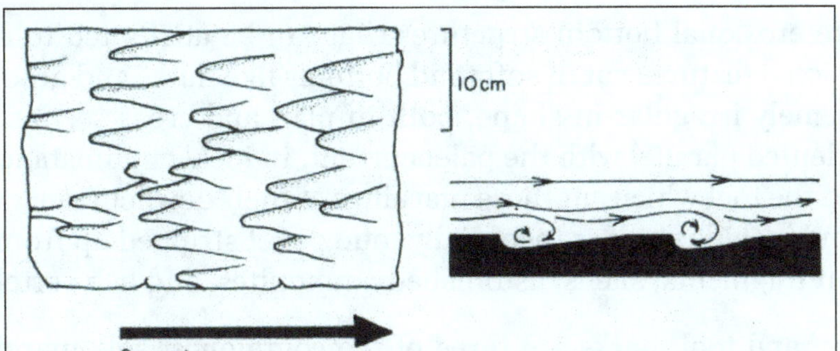

(Left) Sketch of flute marks and (Right) cross-section showing scouring of fluted hollows in soft mud by current vortices.

Photograph of groove marks on the base of a turbidite sand, cut into shale. Hat at top right provides scale (size 8.5). Laingsberg Formation, Ecca Group (Permian), Great Karroo, South Africa.

Groove Marks

The second important type of erosional interbed structure is groove marks. These, like flutes, tend to be cut into mud and overlain by sand. They are long, thin, straight erosional marks. They are seldom more than a few millimeters deep or wide, but they may continue uninterrupted for a meter of more. In cross-section, the grooves are angular or rounded. Grooves occur where sands overlie muds in diverse environmental settings. Like flutes, they are especially characteristic of turbidite sands and trend parallel to the current direction determined from flutes and sedimentary structures within the sands. Grooves are seldom associated with flutes, however, and as they are best developed in a more distal (downcurrent) situation than flutes.

It is clear that grooves are erosional features cut parallel to the current. Their straightness suggests laminar rather than turbulent flow conditions. It has been argued that the grooves are carved by objects borne along in the current. Furthermore, the linearity of the groove proves that the tools were not saltating or rotating, but that they were transported downcurrent at a constant orientation and at an almost constant elevation with respect to the sediment substrate.

Tool Marks

Tool marks are erosional bottom structures that can be attributed to moving clasts. These are erosional features cut in soft mud bottoms like flutes and grooves. They are, however, extremely irregular in shape, both in plan and cross-section, though they are roughly oriented parallel with the paleocurrent. In ideal circumstances it has been possible to find the tool which cut these markings at their downcurrent end. Tools that have been found include pebbles (especially mud pellet stripped up from the bottom), wood and plant fragments, shells, astonished ammonites, and fish vertebrae.

Flutes, grooves, and tool marks are three of the commonest sole markings found as interbed sedimentary structures. All are erosional and all are best seen in, but not exclusive to, turbidite facies. A variety of other sole markings have been described and picturesquely named. Proximal turbidites tend to occur within channels. Moving downcurrent channels give way to tabular basin floor turbidites under which the sequence of erosional bottom structures grades from scour-and-fill to flute marks to groove marks and finally tool marks at their distal end. These changes are related to the downcurrent decrease in flow velocity of any individual turbidite.

Syndepositional (Intrabed) Structures

They are constructional structures that are formed during deposition, thus they are present within sedimentary beds.

Beds

Beds are tabular or lenticular layer of sedimentary rock that can be distinguished from strata above and below by difference in color, composition and texture. Bedding (Bed) is thicker than 1 cm whereas lamination (Lamina) is thinner than 1 cm. Bedding is composed of beds lamination is composed of lamina.

A. Terms used for describing the thickness of beds and laminae B. Field photo of Massive bedding.

Beds also vary in their shape and definition so that planar, wavy and curved types are recognised, and these may be parallel to each other, non-parallel or discontinuous.

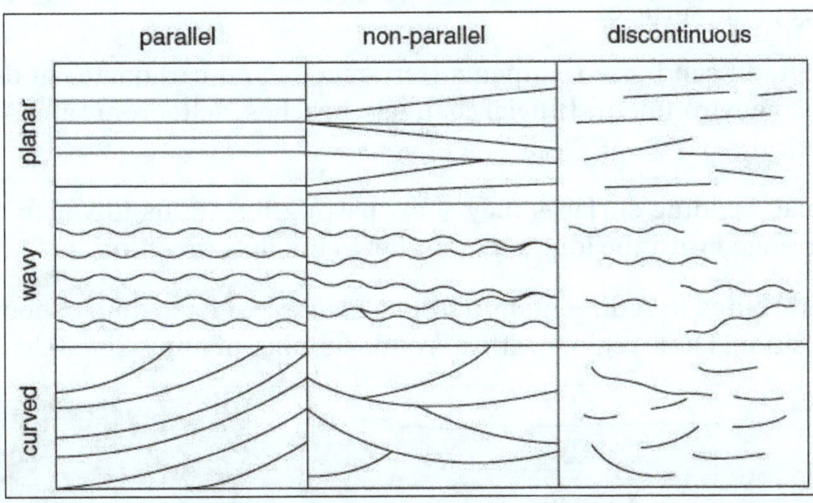

Different types of bedding or lamination.

Bedding Planes or Bedding Surfaces

Bedding planes or bedding surfaces are primary surface in a sedimentary rock separating the upper and lower surfaces of beds, which represent period of nondeposition, erosion, or changes in composition (reflect changes in depositional conditions). Some bedding surfaces may be post-depositional features formed by diagenetic processes.

1. Massive Bedding are beds that appear homogenous and lacking internal sedimentary structures.

Causes of massive bedding:

- Diagenesis: This is particularly characteristic of extensively recrystallized limestones and dolomites.

- Bioturbation: Primary sedimentary structures may be completely destroyed by intensive organic burrowing.
- Primary features (Genuine Massive Bedding):
    - Deposition in quite, low-energy environment (such as some claystones, marls and chalks).
    - Reef rock also commonly lacks bedding.
    - Deposits of mud flows (debris flows), grain flows, and the lower (A unit) part of turbidites (very rapid deposition).

In sandstones massive bedding is rare. It is most frequently seen in very wellsorted sands, where sedimentary structures cannot be delineated by textural variations.

2. Flat-bedding (planar bedding) is planar beds that are deposited horizontally parallel to the bedding plane surfaces.

- It occurs in sand-size sediments (terrigenous and carbonate) in diverse sedimentary environments (fluvial channels, beaches, delta fronts and Aeolian sand sheets).
- Internal, bedding surfaces may show parting lineations in which sand grains are arranged with the long axes parallel to the flow direction.
- It is attributed to sedimentation from a planar bed form under shooting flow or a transitional flow regime with a Froude number of approximately.

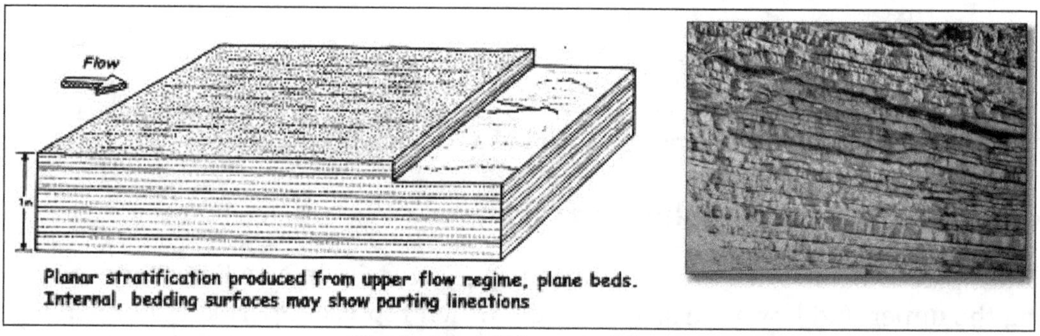

Flat-bedding (planar bedding).

3. Graded bedding is progressive (a gradual) change in grain size from bottom to top of a bed.

- Graded beds range in thickness from a few Cm. to a few meters and commonly have sharp basal contact.
- It is commonly occur at the basal part of turbidities associated with flute and groove marks.

- It is commonly produced by turbidity currents as sediment settles out of suspension, normally during the waning phase of a turbidity flow.
- It helps to determine the base and the top of a bed.

Graded bedding: (Left) Normal grading. (Right) Compound grading.

Types of graded bedding:

- Normal grading: A gradual upward decrease in grain size through a bed (common). It is commonly occur at the basal part of turbidites.
- Coarse tail-grading: A gradual upward decrease in the size of coarse grains through a bed.
- Reverse grading: When the grain size increase upward through a bed (not common).
- Compound (multiple) grading: Where there are several grade subunit in a bed occur.

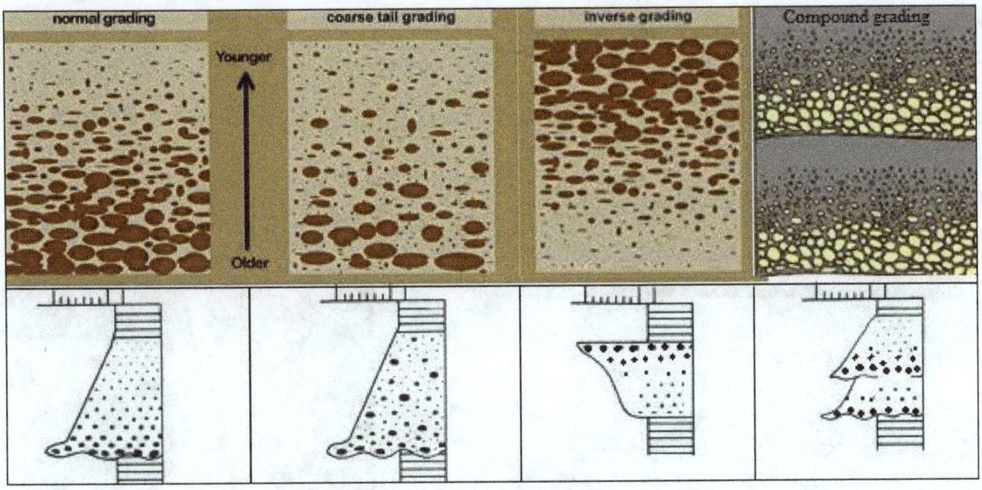

Types of graded bedding.

4. Cross-bedding: It is one of the most common and most important of all sedimentary

structures (the most important paleocurrent indicators). Cross-bedding, consists of inclined dipping bedding, bounded by sub-horizontal surfaces.

- It is ubiquitous in traction current deposits in diverse environments.
- Individual beds range in thickness from a few tens of centimeters to a meter or more.
- It is used to determine current flow directions in ancient rocks (very excellent paleocurrent indicators).
- Each of cross bedded units (Isolated cross-bed) is termed a set. Vertically contiguous sets are termed cosets. The inclined bedding is referred to as a foreset. Foresets may grade down with decreasing dip angle into a bottomset or toeset.

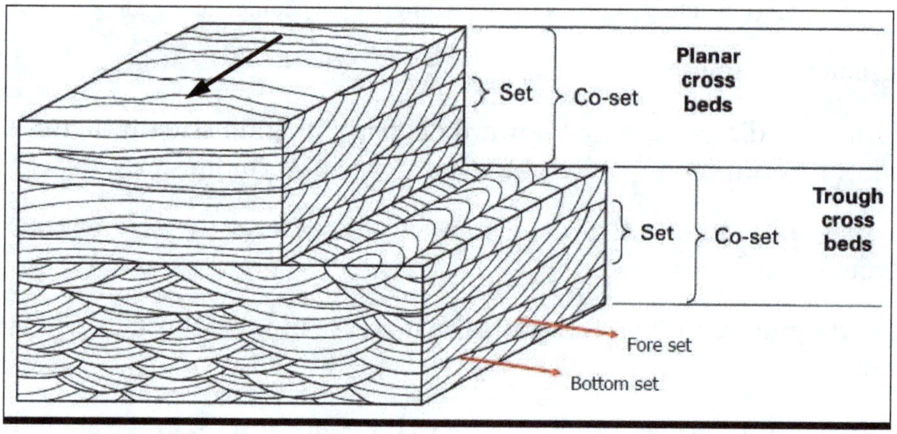

Basic nomenclature of cross-bedding. Tabular planar cross-beds have subplanar foresets. Trough cross-beds have spoon-shaped foresets. Isolated cross-beds are referred to as sets. Vertically grouped foresets constitute a coset.

Formation of cross bedding: Formed mainly by the migration of the crested dunes and megaripples.

- Migration of dune during lower flow regime conditions.

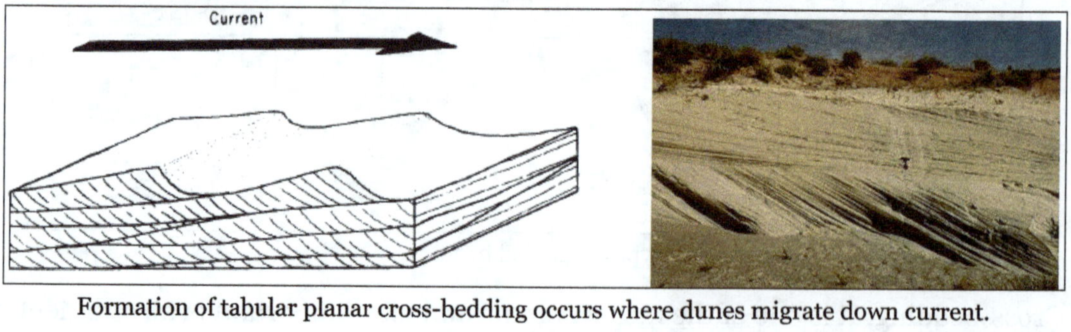

Formation of tabular planar cross-bedding occurs where dunes migrate down current.

- Migration of antidunes in upper flow regime conditions to deposit up current dipping cross-beds.
- In river channels (braided type) a single set of cross-bedded strata deposited in the steep side of obstacles.
- Channel may be infilled by cross-bedding paralleling the channel margin.

Cross-beds are divided on the basis of:

- Overall geometry (foresets shape).
- The nature of the bounding surfaces of the cross-bedded units.

They are divided into:

- Tabular cross-bedding consists of cross-bedded units bounded above and below by planar bounding surfaces. The foreset laminae are also commonly planar, but are curved at the basal surface.
- Wedge-shaped cross bedding consists of cross-bedded units having oblique bounding surfaces and foreset laminae.
- Trough cross-bedding consists of cross-bedded units having upward concave foreset lie within erosion scours which are elongated parallel to current flow, closed up current and truncated down current by further troughs.

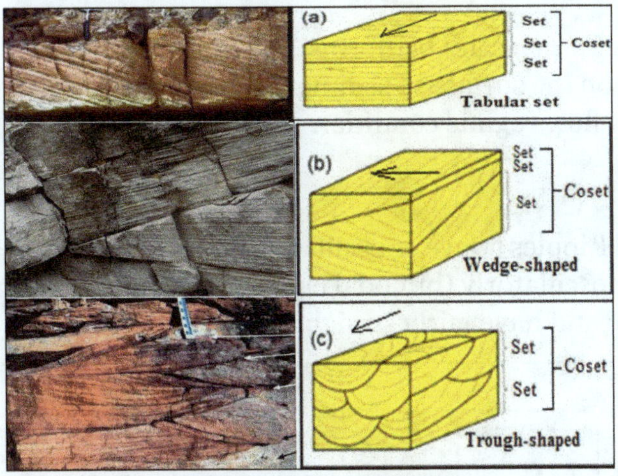

Types of cross bedding.

- Hummocky cross-stratification (HCS) occur typically in fine-grained sandstone and characterized by undulating sets of cross-lamination that are concaveup (swales) and convex-up (hummocks). These cross-beds gently cut into each other with curved erosional surfaces.
    - They form by storm waves in shallow-water, storm-dominated

environments (transition zone between fair-weather wave-base and storm wave-base).

- They lack a specific orientation.
- Hummocky cross-stratified sequences are storm deposits (Tempestites).

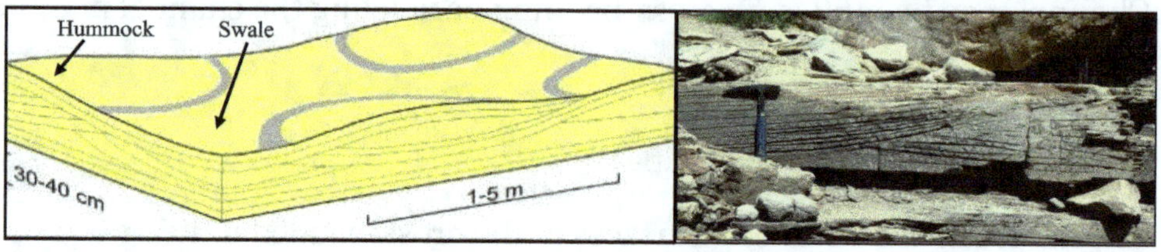

Hummocky cross-stratification (HCS).

- Herring-bone cross bedding is a type of cross-stratification formed in tidal area (tidal channels), where current periodically flow in the opposite direction which reflect the movement of ebb and flood currents.

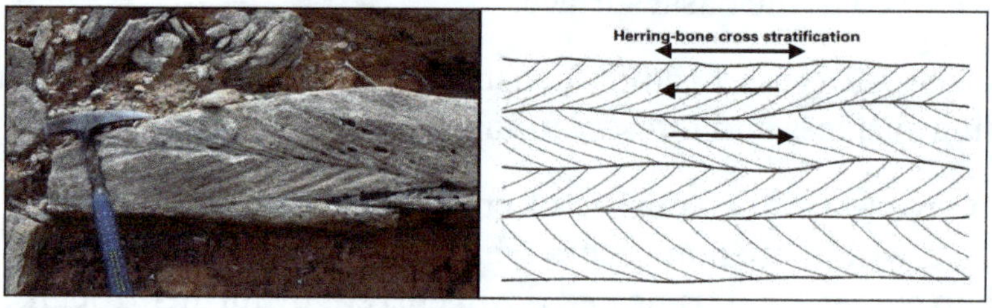

Herring-bone cross bedding.

5. Ripples and dune (large scale) bedforms: Ripples are small-scale alternating ridges and troughs formed on the surface of sediment layer by moving water or wind. They are generated under low flow regime condition (F<1).

Types of ripples:

- Symmetrical Ripples (wave or oscillation ripples) form in shallow water by wave action under oscillatorly (bimodal current) flow and tend to be nearly symmetrical in shape and have fairly straight crests.

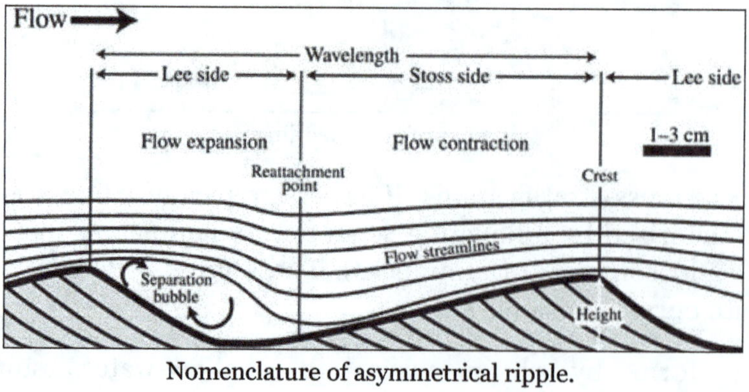

Nomenclature of asymmetrical ripple.

- Asymmetrical ripples (current ripples) formed by wind or water flowing in one direction (unidirectional current). They are asymmetrical in shape, and the steep or lee side faces downstream in the direction of current flow.

When ripples migrate on unconsolidated noncohesive sediment with simultaneous upward growth forming climbing ripples with cross lamination.

In plain view, the crests of current ripples have a variety of shapes: straight, sinuous, catenary, linguoid, and lunate. The plan-view shape of ripples is apparently related to water depth and velocity.

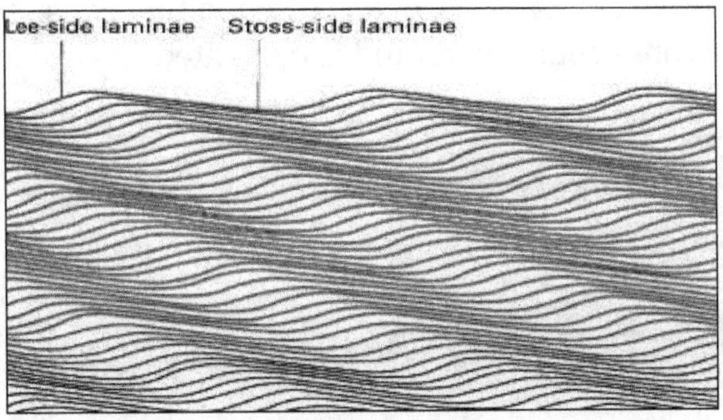

Climbing ripples cross-lamination.

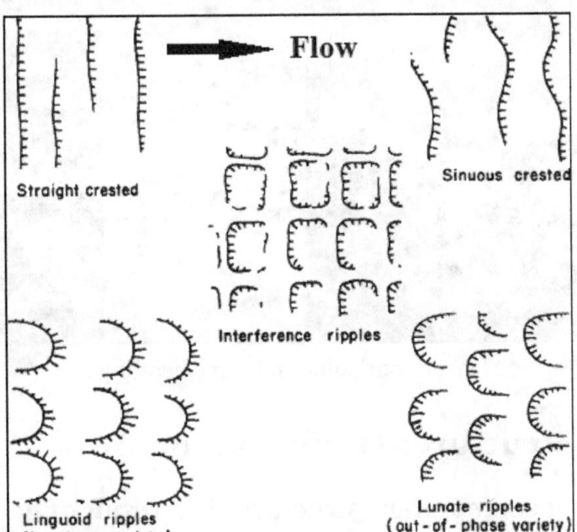

Nomenclature of rippled bed forms in plain view.

Types of ripples bedding (ripples bedforms):

Ripples bedding is characterized by alternating sand and mud laminae or beds including; flaser, lenticular and wavy bedding.

- Flaser bedding is where cross-laminated sand contains mud streaks, usually in the ripple troughs (sand more than mud). These are commonly forms in relatively high energy environments (sand flats).

- Lenticular bedding is a structure formed is where mud dominates and the cross-laminated sand occurs in lenses (mud more than sand). They are commonly forms in relatively low energy environments.
- Wavy bedding is a structure formed by interbedded mud and ripple crosslaminated sand in which the ripples or sand lenses are continuous in both a vertical and a horizontal direction. These are commonly forms in environments that alternate frequently from higher to lower energies (mixed flats).

Flaser and lenticular bedding form particularly on tidal flats and in subtidal environments where conditions of current flow or wave action that cause sand deposition alternate with slack-water conditions when mud is deposited.

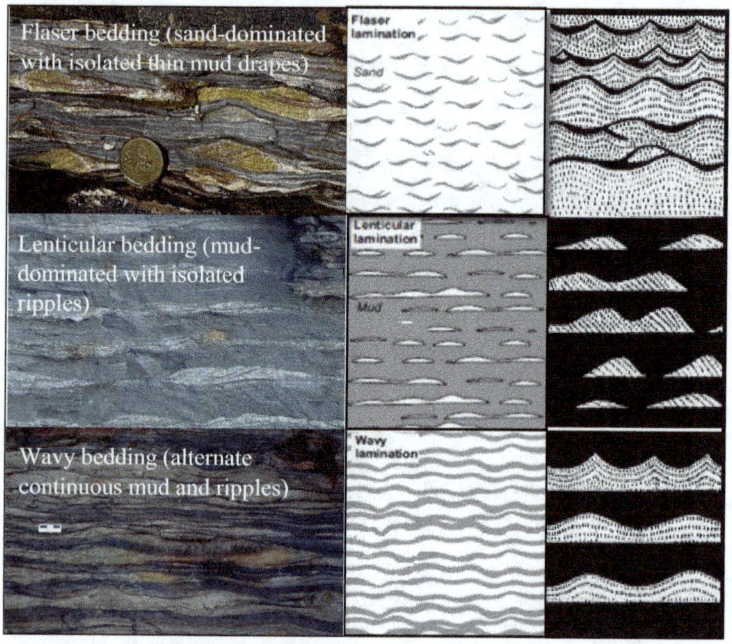

Flaser, lenticular and wavy bedding.

Postdepositional Structures (Deformational)

The third main group of sedimentary structures is a result of deformation. These may be termed postdepositional because, obviously, they can only form after sediment has been laid down.

Table: Classification of Postdepositional Deformational Sedimentary Structures.

Sense of movement	Structure	Nature of deformation
Dominantly vertical	Load casts,Dish structure, and Ball and Pillow structures, Convolute bedding, recumbent foresets, convolute lamination	Plastic (sediments lack shear strength)

Dominantly horizontal	Slumps Slides	Brittle (sediments possess shear strength)

Vertical Plastic Deformational Structures

Deformational structures that involve vertical plastic movement of sediment are of two main types. One group occurs within sand beds and may be loosely referred to as quicksand structures. Structures of the second group develop at the interfaces of sand overlying mud.

A variety of structures develop where sands overlie muds. The mud: sand interface is often deformed in various ways. Most typically irregular-rounded balls of sand depend from the parent sand bed into the mud beneath. These structures are variously termed load casts, ball and pillow structures, etc. They are a variety of the broad group of structures termed sole markings or bottom structures.

Load Structure: Load Cast Generation

- Sole marking generally preserved on the lower side of the sand layer overlying the mud layer.
- Often associated with turbidities with a thin layer of coarser sediment on the top.
- Can also be defined as a bulbous depression formed on the base of a bed of sediment.
- Developed by the differential sinking of the sediment, while still soft, into less dense sediment below.

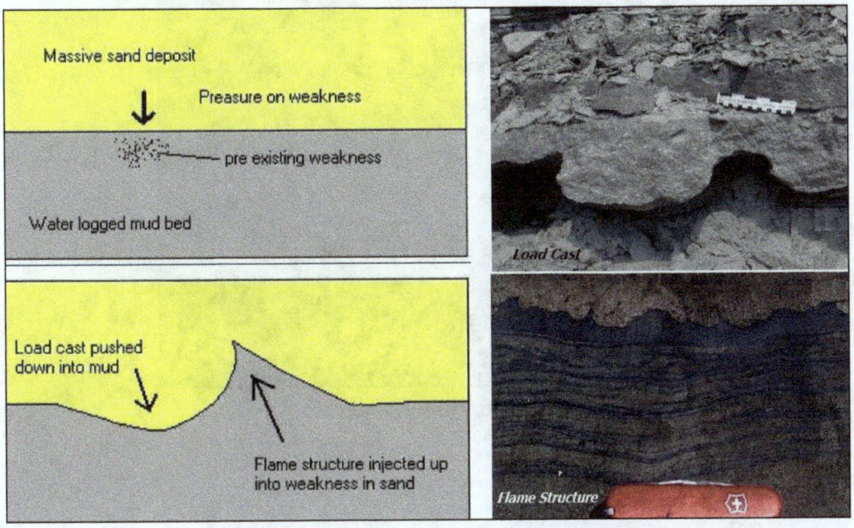

Load Structure: Load Cast & Flame Structure.

Dish Structure

These consist of concave-up laminae, generally a few centimetres across, which may be separated by structureless zones (the pillars). Dish structures and dish-and-pillar structures are formed by the lateral and upward passage of water through sediment.

Dish structure is a particular variant of intra sand deformation. This is seen where lamina or bedding planes are intermittently disrupted and upturned like the rim of a dish. Dish structure is a type of dewatering phenomenon that is particularly characteristic of fluidized sand beds.

Dish structure.

Ball and Pillow Structure

As a result of loading, a bed, usually of sand, can sink into an underlying mud and break up into discrete masses, forming the so-called ball-and-pillow structure.

Ball and Pillow structure.

Convolute Bedding

Quicksand structure involves the deformation of whole beds of sand up to a meter or more thick. It is loosely referred to as convolute bedding. On a smaller scale, laminated fine sands and silts also show penecontemporaneous vertical deformation structures termed convolute lamination. This is similar in geometry to convolute bedding, but occurs in finer grained sediment on a much smaller scale; generally in beds only a decimetre or so high.

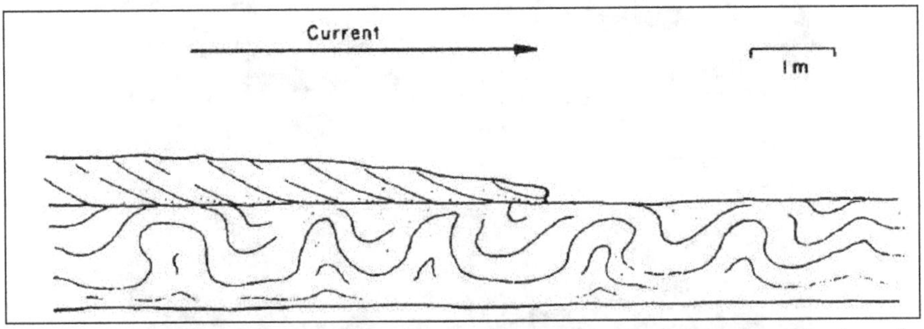

Convolute bedding due to the expulsion of pore water from loosely packed sand.

Recumbent Foreset

These structures are caused by the vertical passage of water through loosely packed sand. This water may be due to a hydrostatic head of water, for example, as is seen on an alluvial fan. Alternatively, the water may be derived from the sediment itself. Sand will not compact significantly at the surface, but its grains may be caused to fall into a tighter packing. This results in a decrease in porosity.

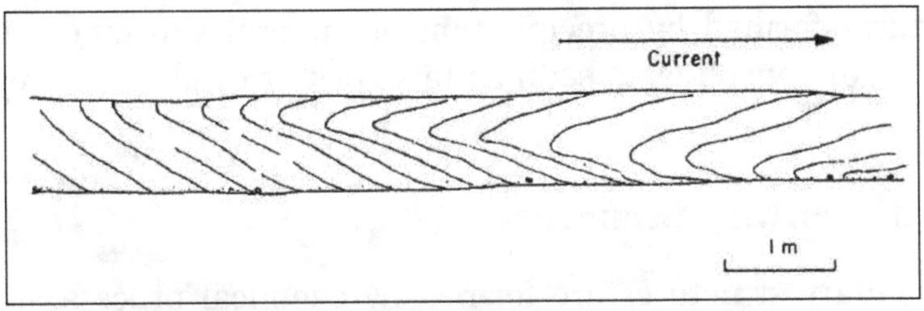

Recumbent foreset deformation in fluvial sandstone.

Convolute Lamination

Laminated fine sands and silts also show penecontemporaneous vertical deformation structures termed convolute lamination. This is similar in geometry to convolute bedding, but occurs in finer grained sediment on a much smaller scale; generally in beds only a decimetre or so high. Convolute lamination is especially characteristic of turbidites, involving deformation of both the laminated and cross-laminated Bouma units.

Horizontal Deformational Structures

Slumps and Slides

Slump structures, like the structures, involve the penecontemporaneous plastic deformation of sand and mud. Slump folds, however, commonly show clear evidence of extensive lateral movement in a consistent direction. Slump folds are commonly associated with penecontemporaneous faulting and with major low angle zones of decollement termed "slide planes."

Slump structure.

Secondary Sedimentary Structures

They are structures formed by precipitation of mineral substances in the pores of semi-consolidated or consolidated sedimentary rock or by chemical replacement processes.

Chemical Sedimentary Structures

Chemical sedimentary structures are formed by chemical processes, such as oxidation-reduction, precipitation, evaporation, etc.

Cone in Cone Structures

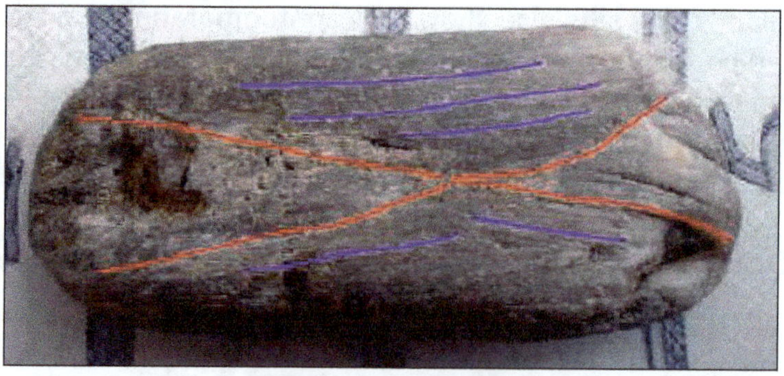

Cone in cone structure.

It is a structure commonly found in fibrous calcite and fibrous gypsum sediments. It has the appearance of a series of cones packed the one inside the other. This structure appears to be due to pressure.

Stylolites

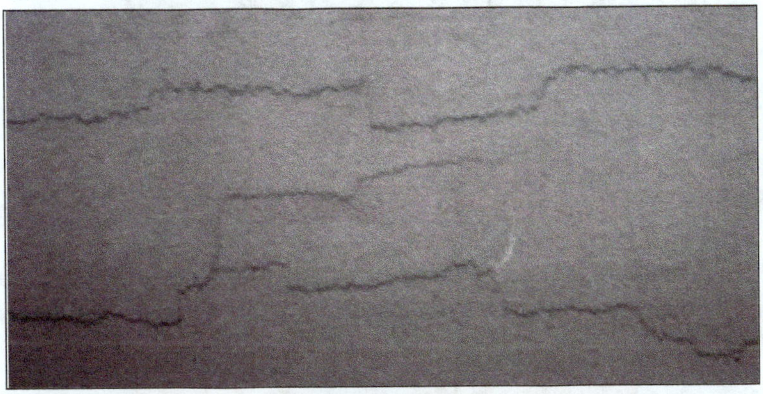

Stylolite structure.

It is formed by the erosional activity of ground water. When the CO_2 rich water percolate through the bedding plane of L.st rock, soluble parts are removed and away and insoluble parts exist there. They appear as irregular structure zig zag line of junction between them. This type of structure is called Stylolites.

Concretionary Structures

Concretionary Structure.

Concretions consist of round or irregular masses of more resistant rock formed as a result of cement precipitating around a core material, usually a fossil or grain of a different composition.

There are two types of Concretionary Structures.

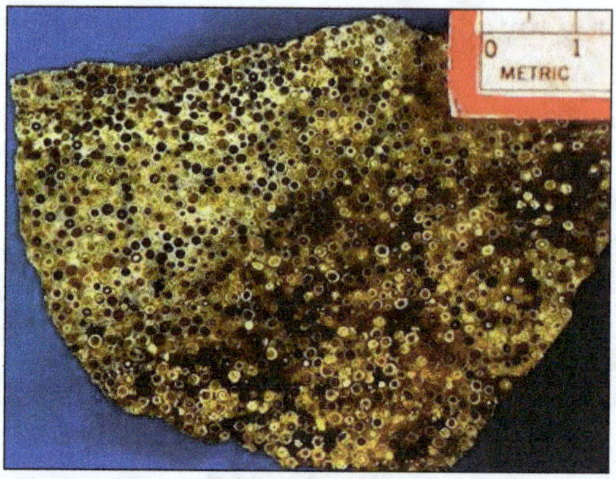

Oolitic structure.

- Oolitic Structures: Oolites are small (0.1-1mm), concentrically layered, spherical grains composed of primary carbonate materials or replacement phases.

Pisolitic structure in Bauxite.

- Pisolitic Structures: Individual size of a concretion is like that of a peanut. Lime stone and Bauxite shows both structure.

Geode

Geode structure.

It is a structure formed by the depositional activity of ground water. It is a cavity in rocks, lined with well-developed crystals of quartz projecting towards the centre.

Vugh

It is a cavity in a rock usually with lining of crystalline minerals. The term used especially of cavities in mineral veins.

Vugh structure.

Crystal Casts

Crystal casts are formed when a soluble crystal (i.e. salt) dissolves in solution, followed by filling of mud or sand in its place.

Crystal casts.

Biogenic Sedimentary Structures

Biogenic sedimentary structures are features which are formed by biological activity.

Stromatolites

These are biogenic laminated structures which have a great variety of growth forms. They develop through the trapping and binding of carbonate particles by a surficial microbial mat ('algal mat') composed mainly of cyanobacteria (formerly called blue-green algae) and other microbes, and the biochemical precipitation of carbonate. Stromatolites are very common in Precambrian carbonate successions but they also occur in many Phanerozoic limestones, particularly those of peritidal origin.

Stromatolites are finely laminated algal accumulations (>10cm in diameter) that result when algae grow upwards trapping carbonate mud into thin layers.

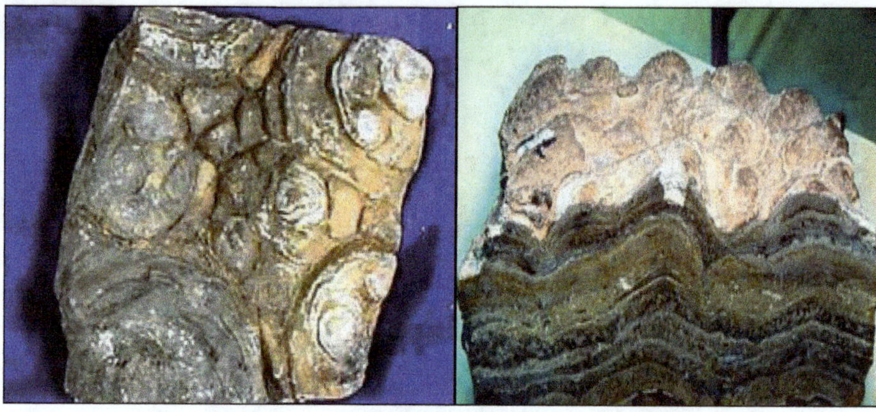

Stromatolite structure.

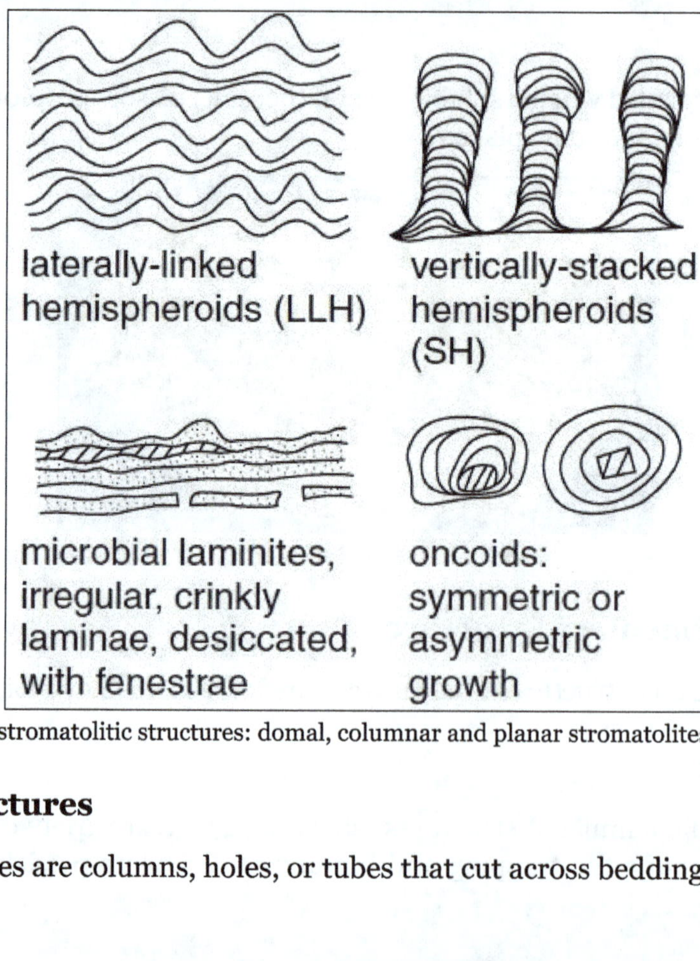

Four common stromatolitic structures: domal, columnar and planar stromatolites, and oncolites.

Escape Structures

Escape structures are columns, holes, or tubes that cut across bedding and are in filled with sediment after the escape of water or a burrowing organism that had been trapped beneath a newly deposited sediment layer.

Escape Structure.

Reefs

Reefs are dome to elongated massive to bedded forms built during carbonate deposition by organisms that biogenetically precipitate carbonate materials.

Reef Structure.

Casts and Moulds

Any depression formed on the bottom of a body of water may become a mold for any sediment that later gets deposited into the depression. The body of sediment that takes on the shape of the mold is referred to as a cast.

Tracks

Tracks or footprints are impressions on the surface of a bed of sediment produced by the feet of animals. Examples include dinosaur footprints or bird tracks. In some cases, tracks are found as sole marks on the bottoms of beds, where sediment has infilled the tracks, and preserved them as casts.

Tracks on the surface of a bed of sediment.

Trails

Trails are groove-like impressions on the surface of a bed of sediment produced by an organism which crawls or drags part of its body. Trails may be straight or curved.

Trails on the surface of a bed of sediment.

Burrows and Borings

Burrows are excavations made by animals into soft sediment. Burrows may be used by organisms for dwellings, or may be produced as a subterranean organism moves through the soil or sediment in search of food.

Burrows are commonly filled in by sediment of a different color or texture than the surrounding sediment, and in some cases, the burrows may have an internally laminated backfilling. Burrow fillings may become cemented and hard, weathering out of the rock in rope-like patterns.

Burrow structure in Triassic rocks.

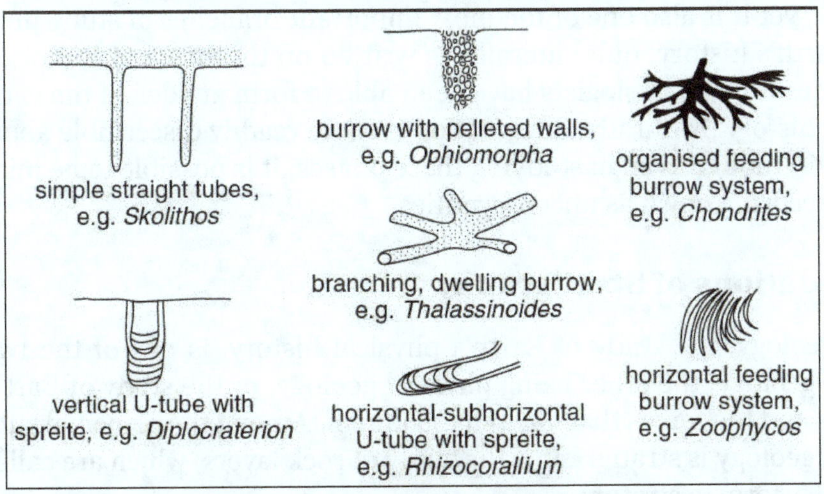

Types of common dwelling and feeding burrows.

Borings are holes made by animals into hard material, such as wood, shells, rock, or hard sediment. Borings are usually circular in cross-section.

References

- Sedimentary-structures, sedimentary-rock, science: britannica.com, Retrieved 02, February 2020
- Sedimentary-Structures-22752660: academia.edu, Retrieved 13, July 2020

Chapter 5

Stratigraphy

The branch of geology which deals with the layers of rocks and layering is called stratigraphy. A few of the major concepts within stratigraphy are successions of strata, stratigraphic nomenclature and Steno's principles of stratigraphy. All the diverse concepts of stratigraphy have been carefully analyzed in this chapter.

Stratigraphy is the study of rock layers (strata) deposited in the earth. It is one of the most challenging of geologic sub-disciplines, comparable to an exacting form of detective work, yet it is also one of the most important branches of study in the geologic sciences. Earth's history, quite literally, is written on the strata of its rocks, and from observing these layers; geologists have been able to form an idea of the various phases in that long history. Naturally, information is more readily discernible about the more recent phases, though even in studying these phases; it is possible to be misled by gaps in the rock record, known as unconformities.

The Foundations of Stratigraphy

Historical geology, the study of Earth's physical history, is one of the two principal branches of geology, the other being physical geology, or the study of Earth's physical components and the forces that have shaped them. Among the principal subdisciplines of historical geology is stratigraphy, the study of rock layers, which are called strata or, in the singular form, a stratum.

Other important subdisciplines include geochronology, the study of Earth's age and the dating of specific formations in terms of geologic time; sedimentology, the study and interpretation of sediments, including sedimentary processes and formations; paleontology, the study of fossilized plants and animals; and paleoecology, the study of the relationship between prehistoric plants and animals and their environments.

Early Work in Stratigraphy

Among the earliest contributions to what could be called historical geology came from the Italian scientist and artist Leonardo da Vinci, who speculated that fossils might have come from the remains of long-dead animals. Nearly two centuries later, stratigraphy itself had its beginnings when the Danish geologist Nicolaus Steno studied the age of rock strata.

Steno formulated what came to be known as the law of superposition, or the idea that strata are deposited in a sequence such that the deeper the layer, the older the rock. This, of course, assumes that the rock has been undisturbed, and it is applicable only for one of the three major types of rock, sedimentary (as opposed to igneous or metamorphic). Later, the German geologist Johann Gottlob Lehmann put forward the theory that certain groups of rocks tend to be associated with each other and that each layer of rock is a sort of chapter in the history of Earth.

Thus, along with Steno, Lehmann helped pioneer the idea of the stratigraphic column, discussed later in this essay. The man credited as the "father of stratigraphy," however, was the English engineer and geologist William Smith. In 1815 Smith produced the first modern geologic map, showing rock strata in England and Wales. Smith's achievement, discussed in Measuring and Mapping Earth, influenced all of geology to the present day by introducing the idea of geologic, as opposed to geographic, mapping. Furthermore, by linking stratigraphy with paleontology, he formulated an important division of stratigraphy, known as biostratigraphy.

Areas of Stratigraphic Study

Along with biostratigraphy, the major areas of stratigraphy include lithostratigraphy, chronostratigraphy, geochronometry, and magnetostratigraphy. The most basic type of stratigraphy, and the first to emerge, was lithostratigraphy, which is simply the study and description of rock layers. Earth scientists working in the area of lithostratigraphy identify various types of layers, which include (from the most specific to the most general), formations, members, beds, groups, and supergroups.

Biostratigraphy involves the study of fossilized plants and animals to establish dates for and correlate relations between stratigraphic layers. Scientists in this field also identify categories of biostratigraphic units, the most basic being a biozone. Magnetostratigraphy is based on the investigation of geomagnetism and the reversals in Earth's magnetic field that have occurred over time.

Chronostratigraphy is devoted to studying the ages of rocks and what they reveal about geologic time, or the vast stretch of history (approximately 4.6 billion years, abbreviated 4.6 Ga) over which Earth's geologic development has occurred. It is concerned primarily with relative dating, whereas geochronometry includes the determination of absolute dates and time intervals. This typically calls for the use of radiometric dating.

Successions of Strata

Think about the sequence of strata you would see in one big outcrop, or in a series of outcrops in a small area. This is called a local section. Even though the physical dimensions of the outcrop might be fairly "equidimensional", you are seeing a stack of strata in essentially one dimension, because the changes from stratum to stratum in the sequence are almost always much more rapid and substantial, per unit distance, than lateral changes along the strata. Figure is a cartoon of an outcrop that shows a local section, along with a schematic column that represents the section you might measure and describe.

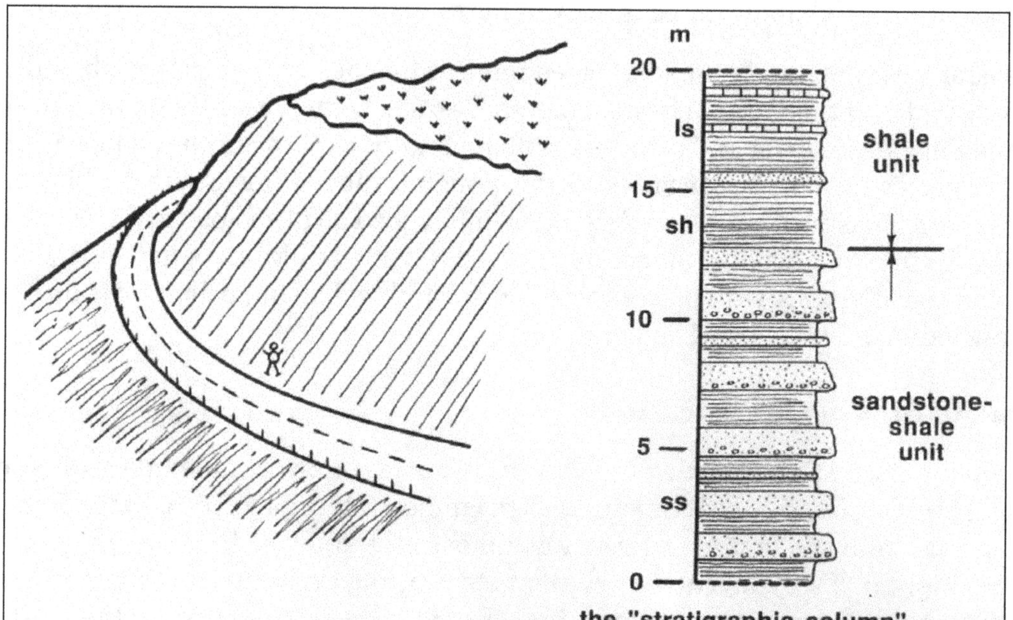

Cartoon of an outcrop showing a local section.

It turns out that in any given area, deposition is usually non-uniform in both space and time; the nature of deposition changes laterally at a given time, and it changes with time at a given point.

It also usually turns out that gradients of deposition in space and time are also non-uniform: at certain times and places, changes are faster and more abrupt. This means that we have a good chance of recognizing what we could think of as natural rock units. When we examine a local vertical section we usually find that vertical variations are not entirely random or chaotic. Instead, there is usually some natural ordering or "signal" such that the stack can be subdivided into rock units that show a high degree of internal sameness compared with rocks above and below.

Contacts between sedimentary rock units can be:

- Gradational by continuous change.
- Gradational by inter bedding.
- Abrupt by change in deposition.
- Abrupt by erosional break.

Now think about what you would see if you traced the beds and rock units laterally—that is, if you hunted up neighboring local sections, maybe kilometers to tens of kilometers away from the highway cut but in the same part of the stratigraphic column.

For example, if we could have seen more of the local section we might have seen the section shown in figure.

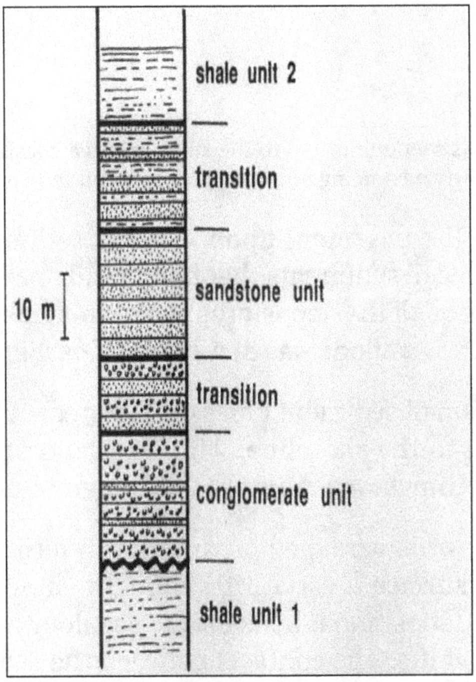

The stratigraphic section of which the section shown in figure is a part.

Because except in unusual depositional settings the individual beds have lateral dimensions smaller than kilometers to tens of kilometers, you should not expect to see exactly the same beds in the neighboring local sections. (Keep in mind that you generally wouldn't really know that, because most beds aren't distinctive enough to be recognized among a whole lot of other similar beds.) But the rock units you recognized at the highway cut probably would be recognizable in neighboring sections.

Your natural inclination would probably to be a "layer-caker" and assume that the times represented by the contacts between rock units are the same from section to section. But you would generally be far wrong in such an assumption! One tip-off would be to find some unique bed in all the sections, like a distinctive bed of volcanic ash, and see where it lies in the section relative to the rock-unit contacts. (Such a bed is called a key bed.) It would generally be in different positions from section to section.

Imagine a coastal area where deposition is taking place and the sea is slowly advancing onto the land (either because sea level is rising around the world, or because the crust is locally subsiding relative to world sea level, or a combination of both). Look at a vertical cross section normal to the shoreline, shown in figure.

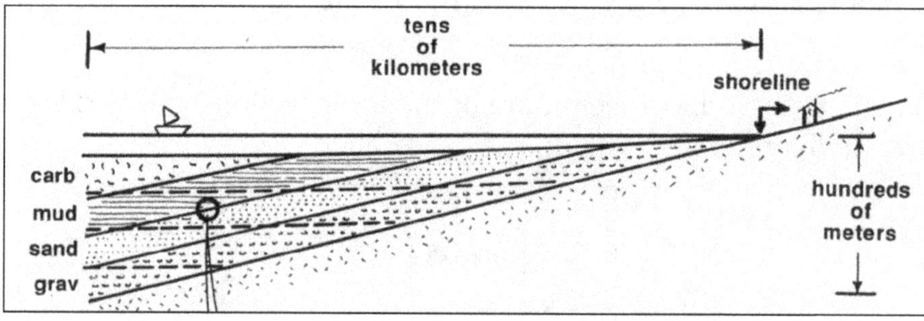

A vertical cross section normal to the shoreline in a coastal area where the sea is slowly advancing onto the land. (Simplified and vertically.)

This cross section shows the basement upon which a sedimentary sequence is being deposited, and the sequence of sediments that has already been deposited on that basement. The dashed lines through the deposit represent imaginary surfaces of equal time; that is, they show where the sea floor was at a series of earlier times.

Look first at the distribution of sediment types on the present sea floor. We have drawn belts of sediment parallel to the shoreline, with finer and finer sediment farther and farther from shore. Often things work this way, although often they don't.

It's striking that the deposit is arranged as units of differing sediment type that are parallel to the basement surface. Just a little thought should convince you that this comes about because the depositional belts on the sea floor are "tied" to the shoreline, and shift as the shoreline shifts. The contacts between the sedimentary units therefore cut across the time surfaces. This nonparallelism of time surfaces and depositional-unit boundaries is termed diachrony.

The sequence of sediment belts on the present sea floor from the shoreline seaward is the same as the sequence of depositional units you would find vertically downward from some point on the sea floor. This is the manifestation of what's called Walther's Law of Facies.

The lateral distribution of facies on a depositional surface is the same as the vertical distribution of these same facies in a vertical section through the deposits. (The term facies, in very common use in sedimentology and stratigraphy, refers to distinctive kinds of sedimentary deposits.)

The advance of the sea onto the land is called a transgression, or marine transgression, and the resulting sedimentary succession is called a transgressive succession.

Now take a closer look at the depositional units in the picture above. Look at one small part of the contact zone between, say, the sand unit and the mud unit, and blow it up to a much larger scale.

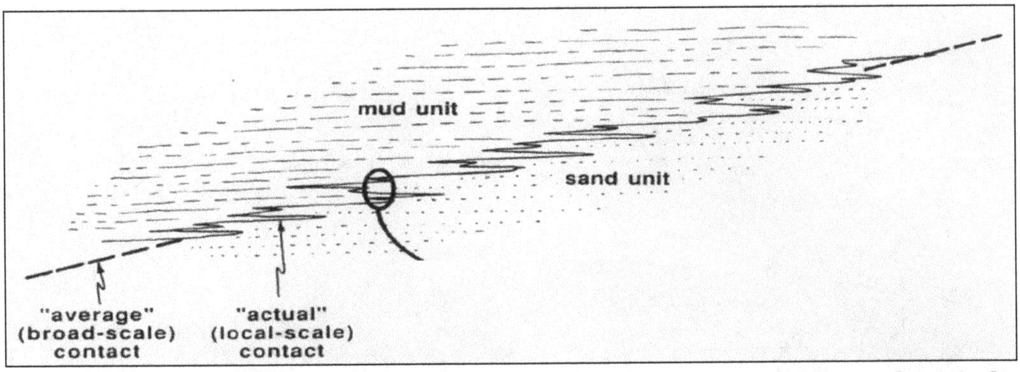

Blown-up view of one small part of the contact zone between the sand unit and the mud unit in figure.

Because of the irregularly shifting nature of the depositional belts over geologically short times (caused by such things as minor fluctuations in sea level superimposed on the general rise, or minor fluctuations in climate, or inherent variability in depositional processes) the boundary would be very jiggly in cross section.

On an even smaller scale, that of individual beds, you would see more jiggliness, caused by such things as storm events.

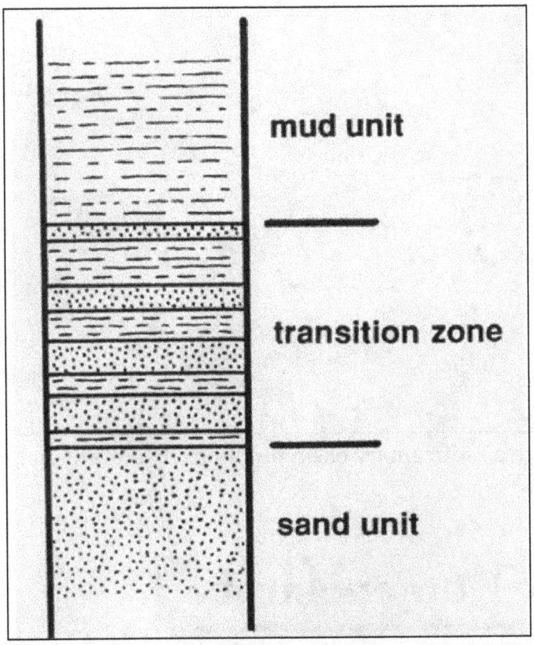

The nature of the contact between the sane unit and the mud unit on an even smaller scale, that of individual beds.

Once this sequence was buried, lithified, uplifted, and eroded for you to see as an outcrop of ancient rock, you would see a succession of interbedded sandstones and shales. The perceptive student might wonder how the picture in figures would differ if sea level was falling and the shoreline was moving seaward. Such a situation is called a marine regression.

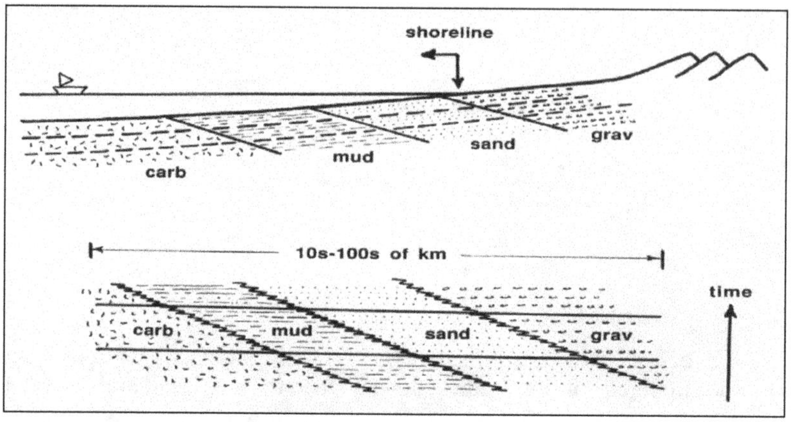

A vertical cross section normal to the shoreline in a coastal area where the sea is slowly retreating from the land. (Simplified and vertically exaggerated.) B) A different way of showing the relationship in Part A, by means of a diagram with equal times represented by horizontal lines.

Finally, the below given figure shows another example of the nonparallelism of time surfaces and depositional-unit boundaries, this time extreme: Filling of a fault trough with sediments shed from the rising block. Here, the contacts of the depositional units cut the time surfaces almost at right angles.

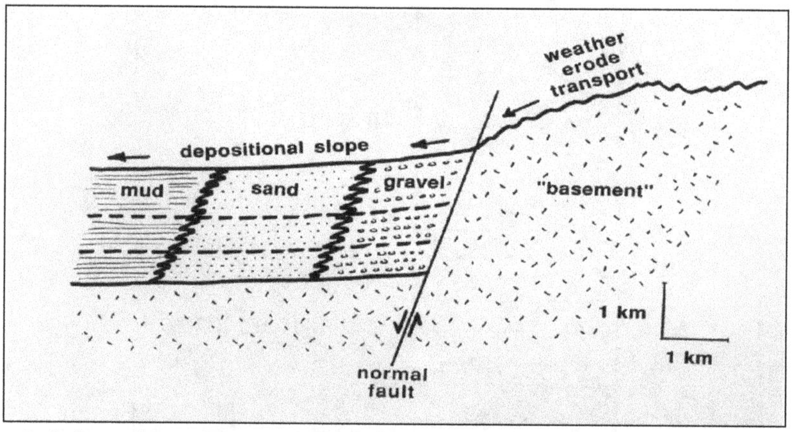

A vertical through a sedimentary basin produced by movement on a normal fault.

Stratigraphic Nomenclature

Lithostratigraphy

Lithostratigraphy is the 'classification of bodies of rock based on the observable lithological properties of the strata and their relative stratigraphic positions. In lithostratigraphy, rock units are considered in terms of the lithological characteristics of the strata and their relative stratigraphic positions. The relative stratigraphic positions of rock units can be determined by considering geometric and physical relationships that indicate which beds are older and which ones are younger. The units can be classified into a hierarchical system of members, formations and groups that provide a basis for categorising and describing rocks in lithostratigraphic terms.

Stratigraphic Relationships

Superposition

Principles of Superposition.

Provided the rocks are the right way up the beds higher in the stratigraphic sequence of deposits will be younger than the lower beds. This rule can be simply applied to a layer-cake stratigraphy but must be applied with care in circumstances where there is a significant depositional topography (e.g. fore-reef deposits may be lower than reef-crest rocks).

Unconformities

An unconformity is a break in sedimentation and where there is erosion of the underlying strata this provides a clear relationship in which the beds below the unconformity are clearly older than those above it. All rocks which lie above the unconformity, or a surface that can be correlated with it, must be younger than those below. In cases where strata have been deformed and partly eroded prior to deposition of the younger beds, an angular unconformity is formed. A disconformity marks a break in sedimentation and some erosion, but without any deformation of the underlying strata.

Cross-cutting Relationships

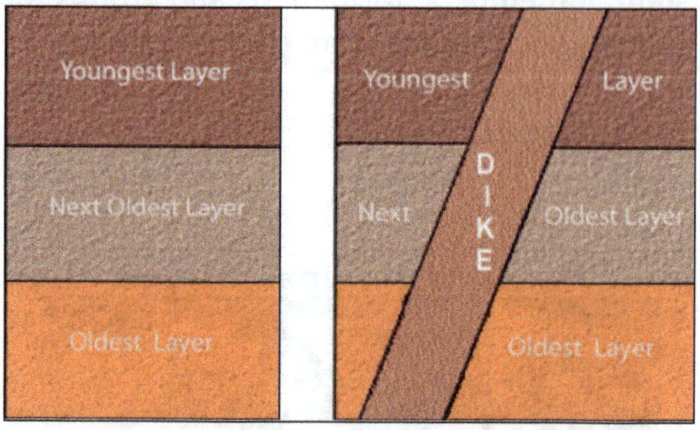

Any unit that has boundaries that cut across other strata must be younger than the rocks it cuts. This is most commonly seen with intrusive bodies such as batholiths on a larger scale and dykes on a smaller scale. This relationship is also seen in fissure fills, sedimentary dykes that form by younger sediments filling a crack or chasm in older rocks.

Included Fragments

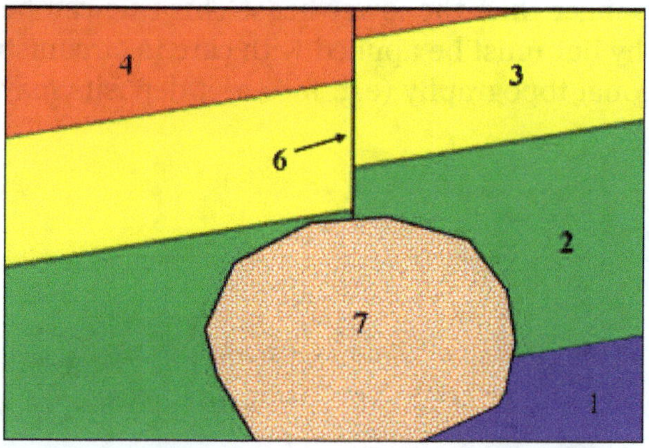

The fragments in a clastic rock must be made up of a rock that is older than the strata in which they are found. The same relationship holds true for igneous rocks that contain pieces of the surrounding country rock as xenoliths. This relationship can be useful in determining the age relationship between rock units that are some distance apart. Pebbles of a characteristic lithology can provide conclusive evidence that the source rock type was being eroded by the time a later unit was being deposited tens or hundreds of kilometres away.

Way-up Indicators in Sedimentary Rocks

The folding and faulting of strata during mountain building can rotate whole successions of beds (formed as horizontal or nearly horizontal layers) through any angle, resulting in beds that may be vertical or completely overturned. In any analysis of deformed strata, it is essential to know the direction of younging, that is, the direction through the layers towards younger rocks. The direction of younging can be determined by small-scale features that indicate the way-up of the beds or by using other stratigraphic techniques to determine the order of formation.

Lithostratigraphic Units

There is a hierarchical framework of terms used for lithostratigraphic units, and from largest to smallest these are: 'Supergroup', 'Group', 'Formation', 'Member' and 'Bed'. The basic unit of lithostratigraphic division of rocks is the formation, which is a body of material that can be identified by its lithological characteristics and by its stratigraphic position. It must be traceable laterally, that is, it must be mappable at the surface or in the subsurface. A formation should have some degree of lithological homogeneity and its defining characteristics may include mineralogical composition, texture, primary sedimentary structures and fossil content in addition to the lithological composition. The material does not necessarily have to be lithified and that all the discussion of terminology and stratigraphic relationships applies equally to unconsolidated sediment. A formation is not defined in terms of its age either by isotopic dating or in terms of biostratigraphy. Information about the fossil content of a mapping unit is useful in the description of a formation but the detailed taxonomy of the fossils that may define the relative age in biostratigraphic terms does not form part of the a lithostratigraphic unit. A formation may be, and often is, a diachronous unit, that is, a deposit with the same lithological properties that was formed at different times in different places. A formation may be divided into smaller units in order to provide more detail of the distribution of lithologies. The term member is used for rock units that have limited lateral extent and are consistently related to a particular formation (or, rarely, more than one formation).

An example would be a formation composed mainly of sandstone but which included beds of conglomerate in some parts of the area of outcrop. A number of members may be defined within a formation (or none at all) and the formation does not have to be completely subdivided in this way: some parts of a formation may not have a member status. Individual beds or sets of beds may be named if they are very distinctive by virtue of their lithology or fossil content. These beds may have economic significance or be useful in correlation because of their easily recognisable characteristics across an area. Where two or more formations are found associated with each other and share certain characteristics they are considered to form a group. Groups are commonly bound by unconformities which can be traced basin-wide. Unconformities that can be identified as major divisions in the stratigraphy over the area of a continent are sometimes considered to be the bounding surfaces of associations of two or more groups known as a supergroup.

Lithology and Characteristics

The field characteristics of the rock, for example, an oolitic grainstone, interbedded coarse siltstone and claystone, a basaltic lithic tuff, and so on form the first part of the description. Although a formation will normally consist mainly of one lithology, combinations of two or more lithologies will often constitute a formation as interbedded or interfingering units. Sedimentary structures (ripple cross-laminations, normal grading, etc.), petrography (often determined from thin-section analysis) and fossil content (both body and trace fossils) should also be noted.

Top and Base

These are the criteria that are used to distinguish beds of this unit from those of underlying and overlying units; this is most commonly a change in lithology from, say, calcareous mudstone to coral boundstone. Where the boundary is not a sharp change from one formation to another, but is gradational, an arbitrary boundary must be placed within the transition. As an example, if the lower formation consists of mainly mudstone with thin sandstone beds, and the upper is mainly sandstone with subordinate mudstone, the boundary may be placed at the point where sandstone first makes up more than 50% of beds. A common convention is for only the base of a unit to be defined at the type section: the top is taken as the defined position of the base of the overlying unit. This convention is used because at another location there may be beds at the top of the lower unit that are not present at the type locality: these can be simply added to the top without a need for redefining the formation boundaries.

Type Section

A type section is the location where the lithological characteristics are clear and, if possible, where the lower and upper boundaries of the formation can be seen. Sometimes it is necessary for a type section to be composite within a type area, with different sections described from different parts of the area. The type section will normally be presented as a graphic sedimentary log and this will form the strato type. It must be precisely located (grid reference and/or GPS location) to make it possible for any other geologist to visit the type section and see the boundaries and the lithological characteristics described.

Thickness and Extent

The thickness is measured in the type section, but variations in the thickness seen at other localities are also noted. The limits of the geographical area over which the unit is recognised should also be determined. There are no formal upper or lower limits to thickness and extent of rock units defined as a formation (or a member or group). The variability of rock types within an area will be the main constraint on the number and thickness of lithostratigraphic units that can be described and defined. Quality and quantity of exposure also play a role, as finer subdivision is possible in areas of good exposure.

Where the age for the formation can be determined by fossil content, radiometric dating or relationships with other rock units this may be included, but this does not form part of the formation. A formation would not be defined as, for example, 'rocks of Burdigalian age', because an interpretation of the fossil content or isotopic dating information is required to determine the age. Information about the facies and interpretation of the environment of deposition might be included but a formation should not be defined in terms of depositional environment, for example, 'lagoonal deposits', as this is an interpretation of the lithological characteristics.

Lithostratigraphic Nomenclature

It helps to avoid confusion if the definition and naming of stratigraphic units follows a set of rules. Formal codes have been set out in publications such as the 'North American Stratigraphic Code' (North American Commission on Stratigraphic Nomenclature 1983) and the 'International Stratigraphic Guide'. The name of the formation, group or member must be taken from a distinct and permanent geographical feature as close as possible to the type section. The lithology is often added to give a complete name such as the Kingston Limestone Formation, but it is not essential or necessarily desirable if the lithological characteristics are varied. The choice of geographical name should be a feature or place marked on topographic maps such as a river, hill, town or village. The rules for naming members, groups and super groups are essentially the same as for formations, but it is not permissible to use a name that is already in use or to use the same name for two different ranks of lithostratigraphic unit. There are some exceptions to these rules of nomenclature that result from historical precedents, and it is less confusing to leave a well-established name as it is rather than to dogmatically revise it. Revisions to stratigraphic nomenclature may become necessary when more detailed work is carried out or more information becomes available. New work in an area may allow a formation to be subdivided and the formation may then be elevated to the rank of group and members may become formations in their own right. For the sake of consistency the geographical name is retained when the rank of the unit is changed.

Lithodemic Units: Non-stratiform Rock Units

The concepts of division into stratigraphic units were developed for rock bodies that are stratiform, layered units, but many metamorphic, igneous plutonic and structurally deformed rocks are not stratiform and they do not follow the rules of superposition. Nonstratiform bodies of rock are called lithodemic units. The basic unit is the lithodeme and this is equivalent in rank to a formation and is also defined on lithological criteria. The word 'lithodeme' is itself rarely used in the name: the body of rock is normally referred to by its geographical name and lithology, such as the White River Granite or Black Hill Schist. An association of lithodemes that share lithological properties, such as a similar metamorphic grade, is referred to as a suite: the term complex is also used as the equivalent to a group for volcanic or tectonically deformed rocks.

Biostratigraphy

Biostratigraphy is the use of fossils in stratigraphy. It relies on the study of in situ fossil distributions to allow recognition of stratigraphically restricted and geographically widespread taxa or populations, which enables subdivision and correlation of lithostratigraphical successions. Such taxa may be selected as index fossils and used as the basis of biostratigraphical correlation – one of the stratigrapher's most powerful tools for correlating Phanerozoic sequences. The basic unit of biostratigraphy is the biozone, which is formally described in terms of its fossil indices and content. Biozones are then ordered in stratigraphical position ultimately to allow correlation of lithostratigraphical units. Biozones can be of any thickness or duration. They can be local to world-wide in scale.

Most Phanerozoic successions with fossil remains, from the lower Cambrian to the present, have been subdivided biostratigraphically. While some of the Lower Palaeozoic zones may be several millions of years in duration, resolution is usually much finer, with a regional resolution of as little as 600 000 years per biozone in the Mesozoic, and considerably less in the Quaternary. Biozones can be further divided. Sub-biozones are defined by taxa that may be of only local importance within the more regionally extensive hierarchy of biozones. Erection of one or more sub-biozones within a biozone does not mean that the whole of the biozone has to be thus subdivided. Biohorizons represent single, laterally widespread palaeontological events, which are not vertically extensive and often provide palaeontological evidence of flooding surfaces or condensed units as applied in sequence stratigraphy.

Biozones were usually defined in areas of supposed 'continuous' sedimentary deposition under marine influences. However, in the Quaternary context, biozones are established for lacustrine and fluvial palaeoenvironments (e.g. using diatoms, spores and pollen, Mollusca, Coleoptera) and terrestrial palaeoenvironments, based on mammalian sequences.

Naming Biozones

Biozones should normally be named after their most characteristic fossil, using standard Linnaean binomial notation (generic form capitalised, specific name lower case, both names italicised, with `biozone' capitalised e.g. Quercus-Pinus Biozone. If a binomial name is used for the biozone, it should be based on a validly published taxon and follow the relevant code of nomenclature. If the name of the taxon then changes, so should the zonal name. Some researchers recommend that the type of biozone should be included in the name (e.g. Palaeoloxodon Taxon-range Biozone), but this can be rather unwieldy.

There are biozonation schemes in common use which do not follow this procedure but instead use alphanumeric notation. These include the Neogene (Tertiary)-Quaternary planktonic foraminiferid schemes and many commercially based biozonations, published or otherwise. Such schemes are too well established to be replaced.

Index Fossils

Suitable fossil indices should be geographically widespread, common, stratigraphically restricted and morphologically distinct enough to enable unambiguous recognition. If possible, the marine taxa selected should be planktonic or nektonic (e.g. dinoflagellates, radiolaria, planktonic foraminifera) to lessen the effects of facies-controlled distribution which limits all living organisms, but which can especially influence the distribution and dispersal of benthic taxa (e.g. benthic foraminifera). Most long-established bizonations have relied on macrofossils (e.g. Mollusca, small and large vertebrates), but drilling activities in both lakes and the sea, as well as in terrestrial settings, have led to the extensive use of microfossil groups. Principal microfossil groups studied extensively over the last 30 years include palynomorphs (spores, pollen, dinoflagellates), foraminifera (planktonic and benthic), nannoplankton, radiolaria, marine diatoms and ostracods.

These microfossil groups possess many of the attributes outlined for biozonal indices, but with two important additions that relate directly to their small size (generally less than 1 mm). Firstly, they are commonly preserved in prodigious numbers (e.g. Globigerina oozes; Radiolarites) and secondly, drilling usually destroys macrofossils but microfossils are normally returned to surface in small rock fragments (ditch cuttings). Conversely, some microfossil taxa are longer ranging than macrofossils so that zonations based on them may be less precise than those based on macrofossils. However, some micropalaeontological, and more typically palynological, schemes for offshore areas are comparable in precision with well-established macrofossil biozonation schemes.

Types of Biozone

There are several categories of biozone, and the biozonation of a single succession can comprise a mixture of types.

Range Biozones reflect the actual range of taxa. A biozonal index does not have to occur throughout the entire vertical extent of its distribution. Hence there are several types of range biozone:

- Total Range Biozone: Boundaries are defined on the combined first (evolution) and last (extinction) occurrences and represent the total stratigraphical and geographical range of the index taxon.
- Local Range Biozone: In practice, it is difficult to define total ranges and many so-called total range bizones are in fact local range biozones.
- Concurrent Range Biozone: Defined on overlapping limits of stratigraphical ranges of two taxa (both tops and bases).

- Partial Range Biozone: Established within the stratigraphical range of a taxon, with the biozone being actually limited by the appearance and/or extinction of other taxa.
- Consecutive Range Biozone: Based on the range of a taxon within an evolving phylogenetic lineage.

Acme and Assemblage biozones are defined on population characteristics:

- Acme Biozone: The boundaries are defined by the maximum extent (abundance/superabundance) of a particular taxon. The subjective nature (i.e. the differing assessments of relative abundance) of acme biozone definition often limits the recognition and regional extent of such units. Additionally, acmes are often related to microhabitat/localized nutrient influxes. Therefore, acme biozones tend not to be applicable over large distances.
- Assemblage Biozone: This is also subjective in definition and is characterized by a distinct assemblage or association of three or more species, which can be distinguished by its character from adjacent assemblages. This type of biozone can be strongly influenced by facies/environment of deposition in certain conditions and can be difficult to define. However, the slow evolution of taxa in the Quaternary means that such biozones are widely recognised.

Boundaries of biozones: Biostratigraphy is a very practical application of the fossil record, so it is hardly surprising that there are two different approaches to the definition of biozonal boundaries.

In 'classical' biostratigraphy, where the biostratigraphical schemes are usually based on macrofossils collected at outcrop, biozones are normally defined by their base, marked by the first appearance (evolutionary or migratory) of a particular fossil (which may not necessarily be the zonal index). The top is then defined by the base of the next overlying unit. In any given section the first appearance may change if subsequent discoveries extend known fossil ranges, so that one cannot apply a 'Golden Spike' approach to biozonal boundaries.

Ranges of individual taxa are often indicated by reference to first (evolutionary) occurrences (FO) and last (evolutionary) occurrences (LO). However, in the commercial world, drilling practices usually do not permit the confident selection of first occurrences (evolutionary inceptions), as downhole contamination of younger sediments (cavings) obscures these events. Therefore, the principal methodology, without recourse to expensive core or sidewall core retrieval, is to delimit biozones based on first downhole occurrences (usually extinction events) rather than evolutionary inceptions.

Regional correlation of a biostratigraphical event may only approximate to a time-line as a consequence of diachronous taxa distribution. This is particularly evident with respect to the distribution and dispersal of benthic populations in marine sequences, which are often facies dependent. It is equally so when organisms respond to climate or environmental change by migration.

Chronostratigraphy and its Units

The element of stratigraphy that deals with the relative time relations and ages of rock bodies.

Chronostratigraphic Classification

The organization of rocks into units on the basis of their age or time of origin. The purpose of chronostratigraphic classification is to organize systematically the rocks forming the Earth's crust into named units (chronostratigraphic units) corresponding to intervals of geologic time (geochronologic units) to serve as a basis for time-correlation and a reference system for recording events of geologic history.

Chronostratigraphic Unit

A body of rocks that includes all rocks formed during a specific interval of geologic time, and only those rocks formed during that time span. Chronostratigraphic units are bounded by synchronous horizons. The rank and relative magnitude of the units in the chronostratigraphic hierarchy are a function of the length of the time interval that their rocks subtend, rather than of their physical thickness.

Chronostratigraphic Horizon (Chronohorizon)

A stratigraphic surface or interface that is synchronous, everywhere of the same age.

Kinds of Chronostratigraphic Units

Hierarchy of formal Chronostratigraphic and Geochronologic Unit terms

Researchers recommend the following formal chronostratigraphic terms and geochronologic equivalents to express units of different rank or time scope.

Table: Conventional Hierarchy of Formal Chronostratigraphic and Geochronologic Terms.

Chronostratigraphic	Geochronologic
Eonothem	Eon
Erathem	Era
System*	Period
Series*	Epoch*

Stage†	Age
Substage	Subage or age

* If additional ranks are needed, the prefixes sub and super may be used with these terms

† Several adjacent stages may be grouped into a superstage.

Position within a chronostratigraphic unit is expressed by adjectives indicative of position such as: basal, lower, middle, upper, etc.; position within a geochronologic unit is expressed by temporal adjectives such as:early, middle, late, etc.

Stage and Age

The stage has been called the basic working unit of chronostratigraphy because it is suited in scope and rank to the practical needs and purposes of intraregional chronostratigraphic classification.

1. Definition: The stage includes all rocks formed during an age. A stage is normally the lowest ranking unit in the chronostratigraphic hierarchy that can be recognized on a global scale. It is a subdivision of a series.

2. Boundaries and stratotypes: A stage is defined by its boundary stratotypes, sections that contain a designated point in a stratigraphic sequence of essentially continuous deposition, preferably marine, chosen for its correlation potential.

The selection of the boundaries of the stages of the Standard Global Chronostratigraphic Scale deserves particular emphasis because such boundaries serve to define not only the stages but also chronostratigraphic units of higher rank, such as series and systems.

3. Time span: The lower and upper boundary stratotypes of a stage represent specific moments in geologic time, and the time interval between them is the time span of the stage. Currently recognized stages vary in time span, but most range between 2 and 10 million years. The thickness of the strata in a stage and its duration in time are independent variables of widely varying magnitudes.

4. Name: The name of a stage should be derived from a geographic feature in the vicinity of its stratotype or type area. In English, the adjectival form of the geographic term is used with an ending in “ian” or “an”. The age takes the same name as the corresponding stage.

Substage and Superstage

A substage is a subdivision of a stage whose equivalent geochronologic term is subage. Adjacent stages may be grouped into a superstage. Names of substages and superstages follow the same rules as those of stages.

Series and Epoch

1. Definition: The series is a chronostratigraphic unit ranking above a stage and below a system. The geochronologic equivalent of a series is an epoch. The terms superseries and subseries have been used only infrequently.

2. Boundaries and boundary-stratotypes: Series are defined by boundary stratotypes.

3. Time span: The time span of currently accepted series ranges from 13 to 35 million years.

4. Name: A new series name should be derived from a geographic feature in the vicinity of its stratotype or type area. The names of most currently recognized series, however, are derived from their position within a system:lower, middle, upper.

Names of geographic origin should preferably be given the ending "ian" or "an".

The epoch corresponding to a series takes the same name as the series except that the terms "lower" and "upper" applied to a series are changed to "early" and "late" when referring to an epoch.

5. Misuse of "series": The use of the term "series" for a lithostratigraphic unit more or less equivalent to a group should be discontinued.

System and Period

1. Definition: A system is a unit of major rank in the conventional chronostratigraphic hierarchy, above a series and below an erathem. The geochronologic equivalent of a system is a period. Occasionally, the terms subsystem and supersystem have been used.

2. Boundaries and boundary-stratotypes: The boundaries of a system are defined by boundary-stratotypes.

3. Time span: The time span of the currently accepted Phanerozoic systems ranges from 30 to 80 million years, except for the Quaternary System that has a time span of only about 1.64 million years.

4. Name: The names of currently recognized systems are of diverse origin inherited from early classifications: some indicate chronologic position (Tertiary, Quaternary), others have lithologic connotation (Carboniferous, Cretaceous), others are tribal (Ordovician, Silurian), and still others are geographic (Devonian, Permian).

Likewise, they bear a variety of endings such as "an", "ic", and "ous". There is no need to standardize the derivation or orthography of the well-established system names. The period takes the same name as the system to which it corresponds.

Erathem and Era

An erathem consists of a group of systems. The geochronologic equivalent of an erathem is an era. The names of erathems were chosen to reflect major changes of the development of life on the Earth: Paleozoic (old life), Mesozoic (intermediate life), and Cenozoic (recent life). Eras carry the same name as their corresponding erathems.

Eonothem and Eon

An eonothem is a chronostratigraphic unit greater than an erathem. The geochronologic equivalent is an eon. Three eonothems are generally recognized, from older to younger, the Archean, Proterozoic and Phanerozoic eonothems. The combined first two are usually referred to as the Precambrian. The eons take the same name as their corresponding eonothems.

Nonhierarchical Formal Chronostratigraphic Units - the Chronozone

1. Definition: A chronozone is a formal chronostratigraphic unit of unspecified rank, not part of the hierarchy of conventional chronostratigraphic units. It is the body of rocks formed anywhere during the time span of some designated stratigraphic unit or geologic feature. The corresponding geochronologic unit is the chron.

2. Time span: The time span of a chronozone is the time span of a previously designated stratigraphic unit or interval, such as a lithostratigraphic, biostratigraphic, or magnetostratigraphic polarity unit. It should be recognized, however, that while the stratigraphic unit on which the chronozone is based extends geographically only as far as its diagnostic properties can be recognized, the corresponding chronozone includes all rocks formed everywhere during the time span represented by the designated unit. For instance, a formal chronozone based on the time span of a biozone includes all strata equivalent in age to the total maximum time span of that biozone regardless of the presence or absence of fossils diagnostic of the biozone.

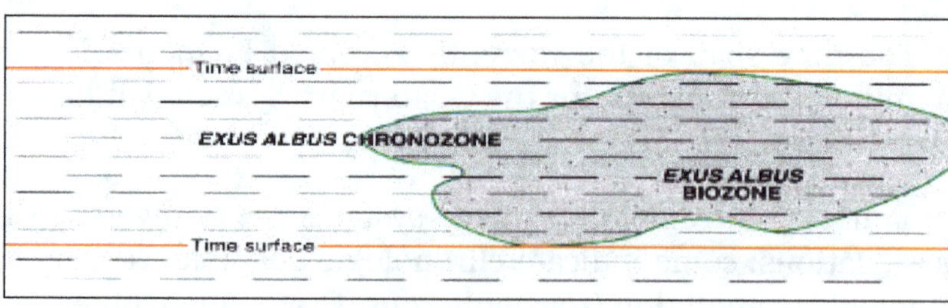

Relation between Exus albus Chronozone and Exus albus Biozone. (Distribution of specimens of Exas albus shown by dot-pattern).

Chronozones may be of widely different time spans. The designation of the boundaries of a chronozone and of its time span can be done in several ways depending on the nature of the stratigraphic unit on which the chronozone is based. If the unit has a designated stratotype, the boundaries and time span of the chronozone can be made to correspond either to those of the unit at its stratotype or to the total time span of the unit, which may be longer than that at the stratotype.

In this second case, the boundaries and time span of the chronozone would vary with increasing information concerning the time span of the unit. If the unit on which the chronozone is based is of the type which cannot appropriately have a designated stratotype, such as a biostratigraphic unit, its time span cannot be defined either because the time span of the reference unit may change with increasing information.

3. Geographic extent: The geographic extent of a chronozone is, in theory, worldwide, but its pplicability is limited to the area over which its time span can be identified, which is usually less.

4. Name: A chronozone takes its name from the stratigraphic unit on which it is based, e.g., Exus albus Chronozone, based on the Exus albus Range Zone.

The Standard Global Chronostratigraphic (Geochronologic) Scale

1. Concept: A major goal of chronostratigraphic classification is the establishment of a hierarchy of chronostratigraphic units of worldwide scope, which will serve as a standard scale of reference for the dating of all rocks everywhere and for relating all rocks everywhere to world geologic history. All units of the standard chronostratigraphic hierarchy are theoretically worldwide in extent, as are their corresponding time spans.

2. Present status: The Standard Global Chronostratigraphic (Geochronologic) Scale can be found in the International Chronostratigraphic Chart.

Regional Chronostratigraphic Scales

The units of the Standard Global Chronostratigraphic (Geochronologic) Scale are valid only as they are based on sound, detailed local and regional stratigraphy. Accordingly, the route toward recognition of uniform global units is by means of local or regional stratigraphic scales. Moreover, regional units will probably always be needed whether or not they can be correlated with the standard global units. It is better to refer strata to local or regional units with accuracy and precision rather than to strain beyond the current limits of time correlation in assigning these strata to units of a global scale. Local or regional chronostratigraphic units are governed by the same rules as are established for the units of the Standard Global Chronostratigraphic Scale.

Subdivision of the Precambrian

The Precambrian has been subdivided into arbitrary geochronometric units, but it has not been subdivided into chronostratigraphic units recognizable on a global scale. There are prospects that chronostratigraphic subdivision of much of the Precambrian may eventually be attained through isotopic dating and through other means of time correlation. However, the basic principles to be used in subdividing the Precambrian into major chronostratigraphic units should be the same as for Phanerozoic rocks, even though different emphasis may be placed on various means of time correlation, predominantly isotopic dating.

Quaternary Chronostratigraphic Units

The basic principles used in subdividing the Quaternary into chronostratigraphic units are the same as for other Phanerozoic chronostratigraphic units, although the methods of time correlation may have a different emphasis. As in the case of other chronostratigraphic units, those of the Quaternary require boundary definitions and designation of boundary stratotypes.

Procedures for Establishing Chronostratigraphic Units

Boundary Stratotypes as Standards

The essential part of the definition of a chronostratigraphic unit is the time span during which the unit described was formed. Since the only record of geologic time and of the events of geologic history lies in the rocks themselves, the best standard for a chronostratigraphic unit is a body of rocks formed between two designated instants of geologic time.

For these reasons, the boundaries of a chronostratigraphic unit of any rank are defined by two designated reference points in the rock sequence. The two points are located in the boundary-stratotypes of the chronostratigraphic unit which need not be part of a single section. Both, however, should be chosen in sequences of essentially continuous deposition since the reference points for the boundaries should represent points in time as specific as possible.

Advantage of Defining Chronostratigraphic Units by their Lower Boundary Stratotypes

The definition of a chronostratigraphic unit places emphasis in the selection of the boundary-stratotype of its lower boundary; its upper boundary is defined as the lower boundary of the succeeding unit. This procedure avoids gaps and overlaps in the Standard Global Chronostratigraphic Scale.

For example, should it be shown that the selected horizon is at the level of an undetected break in the sequence, then the missing span of geologic history would belong to the lower unit by definition and ambiguity is avoided?

Requirements for the Selection of Boundary Stratotypes of Chronostratigraphic Units

Chronostratigraphic units offer the best promise of being identified, accepted, and used globally and of being, therefore, the basis for international communication and understanding because they are defined on the basis of their time of formation, a universal property. Particularly important in this respect are the units of the Standard Global Chronostratigraphic (Geochronologic) Scale. The term "Global Boundary Stratotype Section and Point" (GSSP) has been proposed for the standard boundary-stratotypes of the units of this scale.

In addition to the general requirements for the selection and description of stratotypes, boundary-stratotypes of chronostratigraphic units should fulfill the following requirements:

- The boundary-stratotypes must be selected in sections representing essentially continuous deposition.The worst possible choice for a boundary-stratotype of a chronostratigraphic unit is at an unconformity.
- The boundary-stratotypes of Standard Global Chronostratigraphic Units should be in marine, fossiliferous sections without major vertical lithofacies or biofacies changes. Boundary stratotypes of chronostratigraphic units of local application may need to be in a nonmarine section.
- The fossil content should be abundant, distinctive, well preserved, and represent a fauna and/or flora as cosmopolitan and as diverse as possible.
- The section should be well exposed and in an area of minimal structural deformation or surficial disturbance, metamorphism and diagenetic alteration, and with ample thickness of strata below, above and laterally from the selected boundary-stratotype.
- Boundary stratotypes of the units of the Standard Global Chronostratigraphic Scale should be selected in easily accessible sections that offer reasonable assurance of free study, collection, and long-range preservation. Permanent field markers are desirable.
- The selected section should be well studied and collected and the results of the investigations published, and the fossils collected from the section securely stored and easily accessible for study in a permanent facility.
- The selection of the boundary stratotype, where possible, should take account of historical priority and usage and should approximate traditional boundaries.
- To insure its acceptance and use in the Earth sciences, a boundary stratotype should be selected to contain as many specific marker horizons or other attributes favorable for long-distance time correlation as possible.

Procedures for Extending Chronostratigraphic

The boundaries of chronostratigraphic units are synchronous horizons by definition.

In practice, the boundaries are synchronous only so far as the resolving power of existing methods of time correlation can prove them to be so. All possible lines of evidence should be utilized to extend chronostratigraphic units and their boundaries. Some of the most commonly used are:

Physical Interrelations of Strata

The Law of Superposition states that in an undisturbed sequence of sedimentary strata the uppermost strata are younger than those on which they rest. The determination of the order of superposition provides unequivocal evidence for relative age relations.

All other methods of relative age determination are dependent on the observed physical sequence of strata as a check on their validity. For a sufficiently limited distance, the trace of a bedding plane is the best indicator of synchroneity.

Lithology

Lithologic properties are commonly influenced more strongly by local environment than by age, the boundaries of lithostratigraphic units eventually cut across synchronous surfaces, and similar lithologic features occur repeatedly in the stratigraphic sequence. Even so, a lithostratigraphic unit always has some chronostratigraphic connotation and is useful as an approximate guide to chronostratigraphic position, especially locally. Distinctive and widespread lithologic units also may be diagnostic of chronostratigraphic position.

Paleontology

The orderly and progressive course of organic evolution is irreversible with respect to geologic time and the remains of life are widespread and distinctive. For these reasons, fossil taxa, and particularly their evolutionary sequences, constitute one of the best and most widely used means of tracing and correlating beds and determining their relative age. Biostratigraphic correlation, however, is not time correlation because homotaxy between samples may result from other causes than that the samples are equal in age.

Isotopic Age Determinations

Isotopic dating methods (U-Pb, Rb-Sr, K-Ar, Ar-Ar) based on the radioactive decay of certain parent nuclides at a rate that is constant and suitable for measuring geologic time provide chronostratigraphic data of high precision with analytical errors in the range of 0.1 to 2 percent. However, not all rock types and minerals are amenable to isotopic age determination.

Isotopic dating contributes age values expressed in years and it provides the major hope for working out the ages and age relationships of Precambrian rocks. In some circumstances, isotopic age determinations provide the most accurate or even the only basis for age determination and chronostratigraphic classification of sedimentary, volcanic and other igneous rocks.

Discrepancies in age results may arise from the use of different decay constants. It is important to geological comparisons, therefore, that the uniform sets of decay constants recommended by the IUGS Subcommission on Geochronology be used.

A method of age determination through radioactivity differing from those mentioned above is that based on the proportion of the radiocarbon isotope (^{14}C) to normal carbon in the organic matter of sediments. This method has been extremely valuable but is limited in application to the dating of upper Quaternary strata.

Geomagnetic Polarity Reversals

Periodic reversals of the polarity of the Earth's magnetic field are utilized in chronostratigraphy, particularly in upper Mesozoic and Cenozoic rocks where a magnetic time scale has been developed. Polarity reversals are, however, binary and specific ones cannot be identified without assistance from some other method of dating such as biostratigraphy or isotopic dating.

Paleoclimatic Change

Climatic changes leave imprints on the geological record in the form of glacial deposits, evaporites, red beds, coal deposits, faunal changes, etc.

Their effects on the rocks may be local or widespread and provide valuable information for chronocorrelation, but they must be used in combination with other specific methods.

Paleogeography and Eustatic Changes in Sea Level

As a result of either epeirogenic movements of the land masses or eustatic rises and lowerings of the sea level, certain periods of Earth history are characterized worldwide by a general high or low stand of the continents with respect to sea level. The evidence in the rocks of the resulting transgressions, regressions, and unconformities can furnish an excellent basis for establishing a worldwide chronostratigraphic framework. The identification of a particular event, however, is complicated by local vertical movements and so the method requires auxiliary help in order to identify the events correctly.

Unconformities

Even though a surface of unconformity varies in age and time-value from place to place and is never universal in extent, certain unconformities may serve as useful guides to the approximate placement of chronostratigraphic boundaries. Unconformities, however, cannot fulfill the requirements for the selection of such boundaries.

Orogenies

Crustal disturbances have a recognizable effect on the stratigraphic record. However, the considerable duration of many orogenies, their local rather than worldwide nature, and the difficulty of precise identification make them unsatisfactory indicators of worldwide chronostratigraphic correlation.

Other Indicators

Many other lines of evidence may in some circumstances be helpful as guides to time-correlation and as indicators of chronostratigraphic position. Some are more used than others, but none should be rejected.

Geochronology

"Geochronology means dating of geological events" or "Geochronology is the science of determining the age of rocks, fossils, and sediments using signatures inherent in the rocks themselves".

- Absolute geochronology can be accomplished through radioactive isotopes, whereas relative geochronology is provided by tools such as palaeomagnetism and stable isotope ratios.
- The science of geochronology is the prime tool used in the discipline of chronostratigraphy, which attempts to derive absolute age dates for all fossil assemblages and determine the geologic history of the Earth and extraterrestrial bodies.

Geochronology Methods

- Radiometric dating,
- Fission-track dating,
- Cosmogenic nuclide geochronology,
- Luminescence dating,
- Incremental dating,
- Paleomagnetic dating,
- Magnetostratigraphy,
- Chemostratigraphy,
- Correlation of marker horizons,
- Geological hierarchy of chronological periodization.

Radiometric Dating

- By measuring the amount of radioactive decay of a radioactive isotope with a known half-life, geologists can establish the absolute age of the parent material.
- A number of radioactive isotopes are used for this purpose and depending on the rate of decay are used for dating different geological periods.
- Most radiometric methods are suitable for geological time only, but some such as the radiocarbon method and the $^{40}Ar/^{39}Ar$ dating method can be extended into the time of early human life and into recorded history.

Some of the commonly used techniques are:

- Radiocarbon dating: Measures the decay of carbon-14 in organic material and can be best applied to samples younger than about 60,000 year.
- Uranium–lead (U-Pb) dating: Measures the ratio of two lead isotopes (Pb-206 and Pb-207) to the amount of uranium in a mineral or rock.
- Uranium–thorium (U-Th) dating: This technique is used to date speleothems, corals, carbonates, and fossil bones. Its range is from a few years to about 700,000 years.
- Potassium–argon (K-Ar) dating and argon–argon (Ar-Ar) dating: These techniques date metamorphic, igneous and volcanic rocks.

Fission-track Dating

- Fission track dating is a radiometric dating technique based on analyses of the damage trails, or tracks, left by fission fragments in certain uranium-bearing minerals and glasses.
- It is a relatively simple method of radiometric dating that has made a significant impact on understanding the thermal history of continental crust, the timing of volcanic events, and the source and age of different archeological artifacts.
- Fission tracks are sensitive to heat, and therefore the technique is useful at unravelling the thermal evolution of rocks and minerals.
- Fission track analysis of these minerals provides information about the thermal evolution of the source rocks and therefore can be used to understand provenance and the evolution of mountain belts that shed the sediment.

Fission-track dating is used for:

- Understanding the evolution of mountain belts.

- Determining the source or provenance of sediments.
- Studying the thermal evolution of basins.
- Determining the age of poorly dated strata.
- Dating and provenance determination of archeological artifacts.

Cosmogenic Nuclide Geochronology

- Surface exposure dating or Cosmogenic radionuclide dating is a collection of geochronological techniques for estimating the length of time that a rock has been exposed at or near Earth's surface.
- Surface exposure dating is used to date glacial advances and retreats, erosion history, lava flows, meteorite impacts, rock slides, fault scarps, cave development, and other geological events.
- It is most useful for rocks which have been exposed for between 10 years and 30,000,000 years.

Luminescence Dating

- Luminescence dating techniques observe 'light' emitted from materials such as quartz, diamond, feldspar, and calcite.
- Many types of luminescence techniques are utilized in geology, including Optically Stimulated Luminescence (OSL), Cathodoluminescence (CL), and Thermoluminescence (TL).
- Thermoluminescence and Optically Stimulated Luminescence are used in archaeology to date 'fired' objects such as pottery or cooking stones and can be used to observe sand migration.

Incremental Dating

- Incremental dating techniques allow the construction of year-by-year annual chronologies, which can be temporally fixed (i.e., linked to the present day and thus calendar or sidereal time) or floating.
- Archaeologists use tree-ring dating (dendrochronology) to determine the age of old pieces of wood. Trees usually add growth rings on a yearly basis, with the spacing of rings being wider in high growth years and narrower in low growth years. Patterns in tree-ring growth can be used to establish the age of old wood samples, and also give some hints to local climatic conditions. This technique is useful to about 9,000 years ago for samples from the western United States using overlapping tree-ring series from living and dead wood.

Techniques of Incremental Dating

- Dendrochronology.
- Sampling Ice cores.
- Lichenometry.
- Paleomagnetic dating.
- Varves.
- Speleothems.
- Acanthochronology.
- Sclerochronology.

Paleomagnetic Dating

- A sequence of paleomagnetic poles (usually called virtual geomagnetic poles), which are already well defined in age, constitutes an apparent polar wander path (APWP). Such a path is constructed for a large continental block.
- APWPs for different continents can be used as a reference for newly obtained poles for the rocks with unknown age. For paleomagnetic dating, it is suggested to use the APWP in order to date a pole obtained from rocks or sediments of unknown age by linking the paleopole to the nearest point on the APWP.
- Two methods of paleomagnetic dating have been suggested: (1) the angular method and (2) the rotation method.
- The first method is used for paleomagnetic dating of rocks inside of the same continental block. The second method is used for the folded areas where tectonic rotations are possible.

Magnetostratigraphy

- Magnetostratigraphy is a geophysical correlation technique used to date sedimentary and volcanic sequences.
- The samples are analyzed to determine their characteristic remanent magnetizatio, that is, the polarity of Earth's magnetic field at the time a stratum was deposited.
- This is possible because volcanic flows acquire a thermoremanent magnetization and sediments acquire a depositional remanent magnetization, both of which reflect the direction of the Earth's field at the time of formation.

- This technique is typically used to date sequences that generally lack fossils or interbedded igneous rock.

Techniques

- Polarity Chron.
- Sampling procedures.
- Analytical procedures.
- Correlation and ages.

Chemostratigraphy

- Global trends in isotope compositions, particularly carbon-13 and strontium isotopes, can be used to correlate strata.

Correlation of Marker Horizons

- Marker horizons are stratigraphic units of the same age and of such distinctive composition and appearance, that despite their presence in different geographic sites, there is certainty about their age-equivalence.
- Fossil faunal and floral assemblages, both marine and terrestrial, make for distinctive marker horizons.

Steno's Principles of Stratigraphy

The laws of stratigraphy were pioneered by Danish geologist, Nicholas Steno, who is also considered the father of stratigraphy. Steno established, through observation, that rock layers may not be chaotic as they seem. He concluded that rock layers (strata) are deposited with the upper layers being younger and the lower layers older. This, therefore, informs that rock layers indicate a chronological history of the earth and its past life. Stratigraphic records also preserve major geological events and details pertaining to orogenies (mountain-building), and climatic fluctuation.

As a result, Steno came up with three fundamental principles of stratigraphy, commonly known as Steno's laws.

Principle of Superposition

The principle of superposition, often referred to as the law of superposition is an observation that sedimentary layers of rock at the bottom of a sequence if they undeformed, then they must be older than those at the top. The bottom layers must have been in existence in order to permit the upper layers to have been deposited on top of them.

Layers of Sedimentary Rock Demonstrating the Principle of Superposition

Fossil hunting can be fun but beware of the cliffs. The Church cliffs at Lyme Regis are notoriously unstable and dangerous but they do help to demonstrate the law of superposition.

The law of superposition is regarded as one of the fundamental principles that underpin modern geology, although this principle is very much applicable in other research fields such as archaeology. It helps to provide a basis for the relative dating of strata. As the oldest strata will always be found at the bottom of an undeformed, observable sequence of sedimentary rocks. It is extremely helpful when considering stratigraphical dating, which is governed by the proposition that a layer cannot be older than its constituents.

Principle of Original Horizontality

The Principle of Original Horizontality was proposed by the Danish geological pioneer Nicholas Steno. This principle states that layers of sediment are originally deposited horizontally under the action of gravity. The principle is important to the analysis of folded and tilted strata.

Original Horizontal Strata.

From these observations is derived the conclusion that the Earth has not been static and that great forces have been at work over long periods of time, further leading to the conclusions of the science of plate tectonics; that movement and collisions of large plates of the Earth's crust is the cause of folded strata.

As one of Steno's Laws, the Principle of Original Horizontality served well in the nascent days of geological science. However, it is now known that not all sedimentary layers are deposited purely horizontally.

For instance, coarser grained sediments such as sand may be deposited at angles of up to 15 degrees, held up by the internal friction between grains which prevents them slumping to a lower angle without additional reworking or effort. This is known as the angle of repose, and a prime example is the surface of sand dunes.

The Principle of Lateral Continuity

The principle of lateral continuity states that layers of sediment initially extend laterally in all directions; in other words, they are laterally continuous. As a result, rocks that are otherwise similar, but are now separated by a valley or other erosional feature, can be assumed to be originally continuous.

Layers of sediment do not extend indefinitely; rather, the limits can be recognized and are controlled by the amount and type of sediment available and the size and shape of the sedimentary basin. As long as sediment is transported to an area, it will eventually be deposited. However, as the amount of material lessens away from the source, the layer of that material will become thinner.

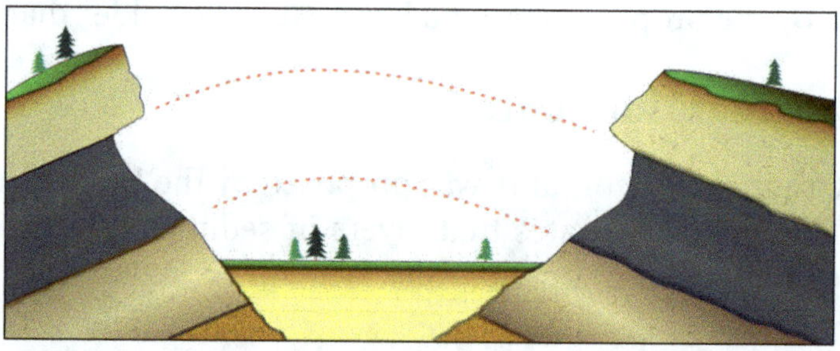

Schematic representation of the principle of lateral continuity.

The Principle of Cross-cutting Relationships

Cross-cutting relationships among geological features have been recognized for many years as one of the fundamental ways of determining relative age relationships between adjacent geological features. The principle of cross-cutting relationships holds that the geological feature which cuts another is the younger of the two features. For example, in the instance of an igneous dike cutting through a layer of sandstone, the dike must be younger than the sandstone.

Cross-cutting relationships are of several basic types. There are structural cross-cutting relationships wherein a fault or fracture cuts through older rock. Stratigraphic cross-cutting relationships occur where an erosional surface (or unconformity) cuts across older rock layers, geological structures, or other geological features. Sedimentologic cross-cutting relationships occur where currents have eroded or scoured older sediment in a local area to produce, for example, a channel filled with sand. Paleontologic cross-cutting relationships occur where animal activity or plant growth produce truncation. This happens, for example, where animal burrows penetrate into pre-existing sedimentary deposits. Geomorphic cross-cutting relationships occur where a surficial feature, such as a river, flows through a gap in a ridge of rock. In a similar example, an impact crater excavates into a subsurface layer of rock.

Cross-cutting relationships may be seen cartographically, megascopically, and microscopically. In other words, these relationships have various scales. A cartographic crosscutting relationship might look like, for example, a large fault dissecting the landscape on a large map. Megascopic crosscutting relationships are features like igneous dikes, which would be seen on an outcrop or in a limited geographic area. Microscopic cross-cutting relationships are those that require study by magnification or other close scrutiny. For example, penetration of a fossil shell by the drilling action of a boring organism is an example of such a relationship.

Cross-cutting relationships may be compound in nature. For example, if a fault were truncated by an unconformity, and that unconformity cut by a dike, we can say, based upon compound cross-cutting relationships that the fault is older than the unconformity and that the unconformity is older than the dike. Using such rationale, the sequence of geological events can be better understood.

Cross-cutting relationships can also be used in conjunction with radiometric age dating to affect an age bracket for geological materials that cannot be directly dated by radiometric techniques. For example, if a layer of sediment containing a fossil of interest is bounded on the top and bottom by unconformities, where the lower unconformity truncates dike A and the upper unconformity truncates dike B (which penetrates the layer in question), this method can be used. A radiometric age date from crystals in dike A will give the maximum age date for the layer in question and likewise, crystals from dike B will give us the minimum age date. This provides an age bracket, or range of possible ages, for the layer in question.

Applications of Stratigraphy

Correlation

The geologist studying the stratigraphic record is a sort of detective, looking for clues. Just as detectives have their methods for solving crimes, geologists rely on correlation, or methods of establishing age relationships between various strata. There are two basic types of correlation: physical correlation, which requires comparison of the physical characteristics of the strata, and fossil correlation, the comparison of fossil types.

Actually, chronostratigraphic work is very similar some of the toughest cases confronted by police detectives, because more often than not the geologic detective has little evidence on which to operate. First of all, only sedimentary rock can be used in making such determinations: for instance, igneous rock in its molten form, as when it is expelled from a volcano, could force itself underneath a rock stratum, thus confusing the stratigraphic record.

Potential Pitfalls

Even when the rock is sedimentary, there is still plenty of room for error. The layers may be many feet or less than an inch deep, and it is up to the geologist to determine whether the stratum has been affected by such geologic forces as erosion. If erosion has occurred, it can cause a disturbance, or unconformity, which tends to render inaccurate any reading of the stratigraphic record.

Another possible source of disturbance is an earthquake, which could cause one part of Earth's crust to shift over an adjacent section, making the stratigraphic record difficult, if not impossible, to read. Under the best of conditions, after all, the strata are hardly neat, easily defined lines. If one observes a horizontal section, there is likely to be a change in thickness, because as the stratum extends outward, it merges with the edges of adjacent deposits.

Yet another potential pitfall in stratigraphic correlation involves one of the most useful tools available to a geologist attempting to find an absolute age for the materials he or she is studying: radiometric dating. Though this method can provide accurate absolute dates, it is quite possible that the age thus determined will be the age of the parent rock from which a sample is taken, not the age of the sample itself. The grains of sand in a piece of sandstone, for instance, are much older than the larger unit of sandstone, and for this reason, radiometric dating is useful only in specific circumstances.

Physical and Fossil Correlation

Physical correlations are achieved on the basis of several criteria, including color, the size of grains, and the varieties of minerals found within a stratum. By such means, it is sometimes possible to correlate widely separated strata.

Particularly impressive feats of correlation can result from the study of fossils, whose stratigraphic implications, as we have noted, were first discovered by William Smith. Smith hit upon the idea of biostratigraphy while excavating land for a set of canals near London. As he discovered, any given stratum contains the same types of fossils, and strata in two different areas thus can be correlated.

Long before Charles Darwin developed the theory of evolution, Smith conceived his own law of faunal succession, which hints at the idea that species developed and disappeared over given phases in Earth's past. According to the law of faunal succession, all samples of any given fossil species were deposited on Earth, regardless of location, at more or less the same time. As a result, if a geologist finds a stratum in one area that contains a particular fossil and another in a distant area containing the same fossil, it is possible to conclude that the strata are the same.

Unconformities

In discussing the many challenges facing a geologist studying stratigraphic data, the role of erosion was noted. Erosion is a source of what are known as unconformities, or gaps in the rock record. Unconformities are of three types: angular unconformities, disconformities, and nonconformities.

Angular unconformities involve a tilting of the layers, such that an upper layer does not lie perfectly parallel to a lower one. Disconformities are more deceptive, because the layers are parallel, yet there is still an unconformity between them, and only a study of the fossil record can reveal the unconformity. Finally, nonconformity arises when sedimentary rocks are divided from a type of igneous rock known as intrusive (meaning "cooled within Earth").

Angular Unconformities

Angular unconformities emerge as a by-product of the dramatic shifts and collisions that take place in plate tectonics. Sediment accumulates and then, as a result of plate movement, is moved about and eventually experiences weathering and erosion. Layers are tilted and then flattened by more erosion, and as the solid earth rises or sinks, they are shifted further. Such is the case, for instance, along the Colorado River at the Grand Canyon, where angular unconformities reveal a series of movements over the years.

Another famous angular unconformity can be found at Siccar Point in Scotland, where nearly horizontal deposits of sandstone rest atop nearly vertical ones of graywacke, another sedimentary rock. Observations of this unconformity led the great geologist James Hutton to the realization that Earth is much, much older than the 6,000 years claimed by theologians in his day.

References

- Sedimentary-geology-spring-2007, earth-atmospheric-and-planetary-sciences: mit.edu, Retrieved 06, March 2020
- Description-of-lithostratigraphy: blogspot.com, Retrieved 14, August 2020
- Biostratigraphy, stratigraphic-guide: stratigraphy.org, Retrieved 22, April 2020
- Chronostratigraphy-age-or-time-of-origin: stratigraphy.org, Retrieved 01, June 2020

- Geochronology: gargicollege.in, Retrieved 10, January 2020
- Basic-principals-of-Stratigraphy-converted-05042020: asutoshcollege.in, Retrieved 13, May 2020
- The-principle-of-superposition: everythingdinosaur.co.uk, Retrieved 21, February 2020
- Principle-of-original-horizontality: geologyin.com, Retrieved 26, July 2020

Chapter 6

Sedimentary Rocks: Diverse Applications

Sedimentary rocks such as limestone and sandstone have a diverse range of applications in a wide variety of industries. A few of these are dimension stone, ceramics, petroleum geology and aquifers. This chapter closely examines these key applications of sedimentary rocks to provide an extensive understanding of the subject.

Ceramic

A ceramic is an inorganic, nonmetallic solid material comprising metal, nonmetal or metalloid atoms primarily held in ionic and covalent bonds.

A Ming Dynasty porcelain vase dated to 1403–1424.

The crystallinity of ceramic materials ranges from highly oriented to semi-crystalline, and often completely amorphous (e.g., glasses). Varying crystallinity and electron consumption in the ionic and covalent bonds cause most ceramic materials to be good thermal and electrical insulators (extensively researched in ceramic engineering). With such a large range of possible options for the composition/structure of a ceramic (e.g. nearly all of the elements, nearly all types of bonding, and all levels of crystallinity), the breadth of the subject is vast, and identifiable attributes (e.g. hardness, toughness, electrical conductivity, etc.) are hard to specify for the group as a whole. General properties such as high melting temperature, high hardness, poor conductivity, high moduli of elasticity, chemical resistance and low ductility are the norm, with known exceptions to each of these rules (e.g. piezoelectric ceramics, glass transition temperature, superconductive ceramics, etc.). Many composites, such as fiberglass and carbon fiber, while containing ceramic materials, are not considered to be part of the ceramic family.

A selection of silicon nitride components.

Mid-16th century ceramic tilework on the Dome of the Rock, Jerusalem.

The earliest ceramics made by humans were pottery objects (i.e. *pots* or *vessels*) or figurines made from clay, either by itself or mixed with other materials like silica, hardened, sintered, in fire. Later ceramics were glazed and fired to create smooth, colored surfaces, decreasing porosity through the use of glassy, amorphous ceramic coatings on top of the crystalline ceramic substrates. Ceramics now include domestic, industrial and building products, as well as a wide range of ceramic art. In the 20th century, new ceramic materials were developed for use in advanced ceramic engineering, such as in semiconductors.

Spherical Hanging Ornament, 1575-1585, Ottoman Period. Brooklyn Museum.

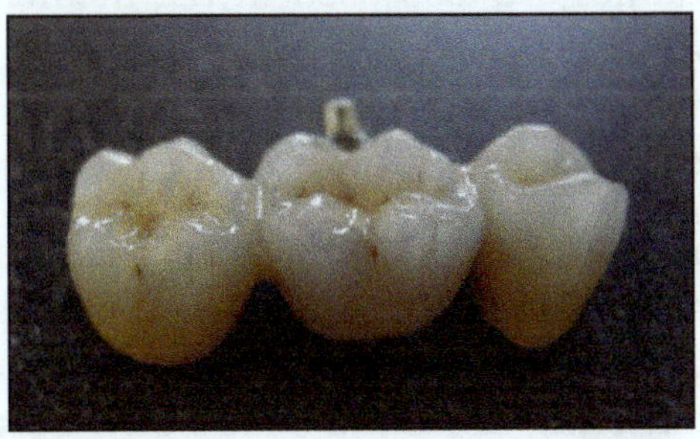

Fixed partial porcelain denture, or "bridge".

The earliest known mention of the root "ceram-" is the "workers of ceramics", written in Linear B syllabic script. The word "ceramic" may be used as an adjective to describe a material, product or process, or it may be used as a noun, either singular, or, more commonly, as the plural noun "ceramics".

Types of Ceramic Material

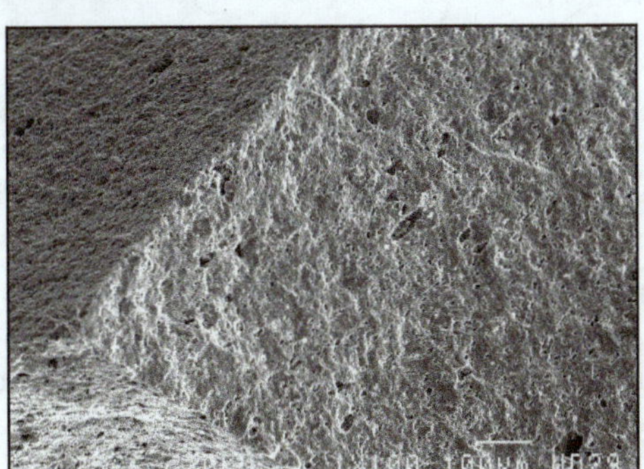

A low magnification SEM micrograph of an advanced ceramic material. The properties of ceramics make fracturing an important inspection method.

A ceramic material is an inorganic, non-metallic, often crystalline oxide, nitride or carbide material. Some elements, such as carbon or silicon, may be considered ceramics.

Ceramic materials are brittle, hard, strong in compression, weak in shearing and tension. They withstand chemical erosion that occurs in other materials subjected to acidic or caustic environments. Ceramics generally can withstand very high temperatures, such as temperatures that range from 1,000 °C to 1,600 °C (1,800 °F to 3,000 °F). Glass is often not considered a ceramic because of its amorphous (noncrystalline) character. However, glassmaking involves several steps of the ceramic process and its mechanical properties are similar to ceramic materials.

Traditional ceramic raw materials include clay minerals such as kaolinite, whereas more recent materials include aluminium oxide, more commonly known as alumina. The modern ceramic materials, which are classified as advanced ceramics, include silicon carbide and tungsten carbide. Both are valued for their abrasion resistance, and hence find use in applications such as the wear plates of crushing equipment in mining operations. Advanced ceramics are also used in the medicine, electrical, electronics industries and body armor.

Crystalline Ceramics

Crystalline ceramic materials are not amenable to a great range of processing. Methods for dealing with them tend to fall into one of two categories – either make the ceramic in the desired shape, by reaction *in situ*, or by "forming" powders into the desired shape, and then sintering to form a solid body. Ceramic forming techniques include shaping by hand (sometimes including a rotation process called "throwing"), slip casting, tape casting (used for making very thin ceramic capacitors, e.g.), injection molding, dry pressing, and other variations. Details of these processes are described in the two books listed below. A few methods use a hybrid between the two approaches.

Noncrystalline Ceramics

Noncrystalline ceramics, being glass, tend to be formed from melts. The glass is shaped when either fully molten, by casting, or when in a state of toffee-like viscosity, by methods such as blowing into a mold. If later heat treatments cause this glass to become partly crystalline, the resulting material is known as a glass-ceramic, widely used as cook-top and also as a glass composite material for nuclear waste disposal.

Properties of Ceramics

The physical properties of any ceramic substance are a direct result of its crystalline structure and chemical composition. Solid state chemistry reveals the fundamental connection between microstructure and properties such as localized density variations, grain size distribution, type of porosity and second-phase content, which can all be correlated with ceramic properties such as mechanical strength σ by the Hall-Petch equation, hardness, toughness, dielectric constant, and the optical properties exhibited by transparent materials.

Physical properties of chemical compounds which provide evidence of chemical composition include odor, colour, volume, density (mass / volume), melting point, boiling point, heat capacity, physical form at room temperature (solid, liquid or gas), hardness, porosity, and index of refraction.

Ceramography is the art and science of preparation, examination and evaluation of ceramic microstructures. Evaluation and characterization of ceramic microstructures is often implemented on similar spatial scales to that used commonly in the emerging field of nanotechnology: from tens of angstroms (A) to tens of micrometers (μm). This is typically somewhere between the minimum wavelength of visible light and the resolution limit of the naked eye.

The microstructure includes most grains, secondary phases, grain boundaries, pores, micro-cracks, structural defects and hardness microindentions. Most bulk mechanical, optical, thermal, electrical and magnetic properties are significantly affected by the observed microstructure. The fabrication method and process conditions are generally indicated by the microstructure. The root cause of many ceramic failures is evident in the cleaved and polished microstructure. Physical properties which constitute the field of materials science and engineering include the following:

Mechanical Properties

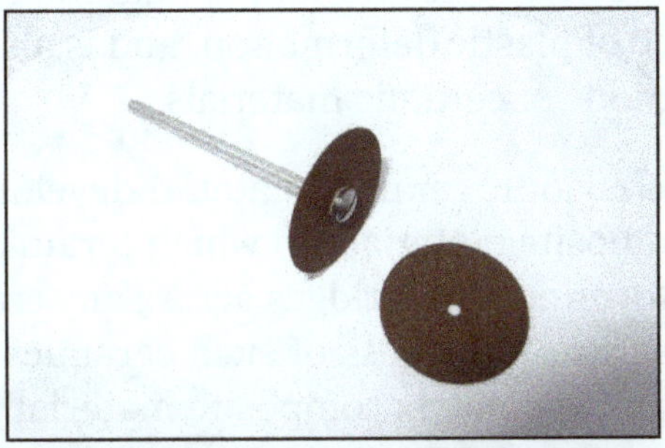

Cutting disks made of silicon carbide.

The Porsche Carrera GT's carbon-ceramic (silicon carbide) disc brake.

Mechanical properties are important in structural and building materials as well as textile fabrics. They include the many properties used to describe the strength of materials such as: elasticity / plasticity, tensile strength, compressive strength, shear strength, fracture toughness & ductility (low in brittle materials), and indentation hardness.

In modern materials science, fracture mechanics is an important tool in improving the mechanical performance of materials and components. It applies the physics of stress and strain, in particular the theories of elasticity and plasticity, to the microscopic crystallographic defects found in real materials in order to predict the macroscopic mechanical failure of bodies. Fractography is widely used with fracture mechanics to understand the causes of failures and also verify the theoretical failure predictions with real life failures.

Ceramic materials are usually ionic or covalent bonded materials, and can be crystalline or amorphous. A material held together by either type of bond will tend to fracture before any plastic deformation takes place, which results in poor toughness in these materials. Additionally, because these materials tend to be porous, the pores and other microscopic imperfections act as stress concentrators, decreasing the toughness further, and reducing the tensile strength. These combine to give catastrophic failures, as opposed to the normally much more gentle failure modes of metals.

These materials do show plastic deformation. However, due to the rigid structure of the crystalline materials, there are very few available slip systems for dislocations to move, and so they deform very slowly. With the non-crystalline (glassy) materials, viscous flow is the dominant source of plastic deformation, and is also very slow. It is therefore neglected in many applications of ceramic materials.

To overcome the brittle behaviour, ceramic material development has introduced the class of ceramic matrix composite materials, in which ceramic fibers are embedded and with specific coatings are forming fiber bridges across any crack. This mechanism substantially increases the fracture toughness of such ceramics. The ceramic disc brakes are, for example using a ceramic matrix composite material manufactured with a specific process.

Electrical Properties

Semiconductors

Some ceramics are semiconductors. Most of these are transition metal oxides that are II-VI semiconductors, such as zinc oxide.

While there are prospects of mass-producing blue LEDs from zinc oxide, ceramicists are most interested in the electrical properties that show grain boundary effects.

One of the most widely used of these is the varistor. These are devices that exhibit the property that resistance drops sharply at a certain threshold voltage. Once the voltage across the device reaches the threshold, there is a breakdown of the electrical structure in the vicinity of the grain boundaries, which results in its electrical resistance dropping from several megohms down to a few hundred ohms. The major advantage of these is that they can dissipate a lot of energy, and they self-reset – after the voltage across the device drops below the threshold, its resistance returns to being high.

This makes them ideal for surge-protection applications; as there is control over the threshold voltage and energy tolerance, they find use in all sorts of applications. The best demonstration of their ability can be found in electrical substations, where they are employed to protect the infrastructure from lightning strikes. They have rapid response, are low maintenance, and do not appreciably degrade from use, making them virtually ideal devices for this application.

Semiconducting ceramics are also employed as gas sensors. When various gases are passed over a polycrystalline ceramic, its electrical resistance changes. With tuning to the possible gas mixtures, very inexpensive devices can be produced.

Superconductivity

The Meissner effect demonstrated by levitating a magnet above a cuprate superconductor, which is cooled by liquid nitrogen.

Under some conditions, such as extremely low temperature, some ceramics exhibit high temperature superconductivity. The exact reason for this is not known, but there are two major families of superconducting ceramics.

Ferroelectricity and Supersets

Piezoelectricity, a link between electrical and mechanical response, is exhibited by a large number of ceramic materials, including the quartz used to measure time in watches and other electronics. Such devices use both properties of piezoelectrics, using electricity to produce a mechanical motion (powering the device) and then using this mechanical motion to produce electricity (generating a signal). The unit of time measured is the natural interval required for electricity to be converted into mechanical energy and back again.

The piezoelectric effect is generally stronger in materials that also exhibit pyroelectricity, and all pyroelectric materials are also piezoelectric. These materials can be used to inter convert between thermal, mechanical, or electrical energy; for instance, after synthesis in a furnace, a pyroelectric crystal allowed to cool under no applied stress generally builds up a static charge of thousands of volts. Such materials are used in motion sensors, where the tiny rise in temperature from a warm body entering the room is enough to produce a measurable voltage in the crystal.

In turn, pyroelectricity is seen most strongly in materials which also display the ferroelectric effect, in which a stable electric dipole can be oriented or reversed by applying an electrostatic field. Pyroelectricity is also a necessary consequence of ferroelectricity. This can be used to store information in ferroelectric capacitors, elements of ferroelectric RAM.

The most common such materials are lead zirconate titanate and barium titanate. Aside from the uses mentioned above, their strong piezoelectric response is exploited in the design of high-frequency loudspeakers, transducers for sonar, and actuators for atomic force and scanning tunneling microscopes.

Positive Thermal Coefficient

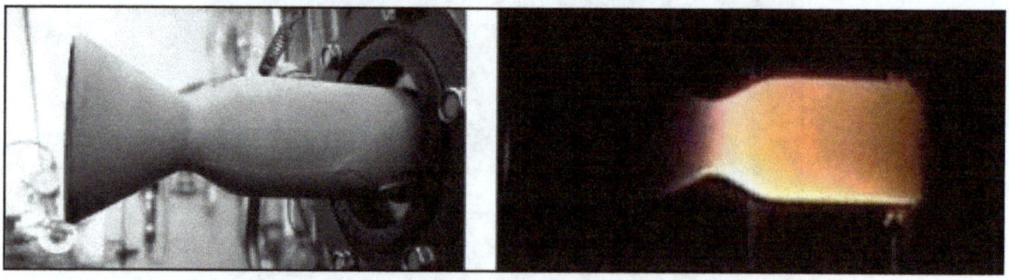

Silicon nitride rocket thruster. Left: Mounted in test stand. Right: Being tested with H_2/O_2 propellants.

Increases in temperature can cause grain boundaries to suddenly become insulating in some semiconducting ceramic materials, mostly mixtures of heavy metal titanates. The critical transition temperature can be adjusted over a wide range by variations in chemistry. In such materials, current will pass through the material until joule heating brings it to the transition temperature, at which point the circuit will be broken and current flow will cease. Such ceramics are used as self-controlled heating elements in, for example, the rear-window defrost circuits of automobiles.

At the transition temperature, the material's dielectric response becomes theoretically infinite. While a lack of temperature control would rule out any practical use of the material near its critical temperature, the dielectric effect remains exceptionally strong even at much higher temperatures. Titanates with critical temperatures far below room temperature have become synonymous with "ceramic" in the context of ceramic capacitors for just this reason.

Optical Properties

Cermax xenon arc lamp with synthetic sapphire output window.

Optically transparent materials focus on the response of a material to incoming lightwaves of a range of wavelengths. Frequency selective optical filters can be utilized to alter or enhance the brightness and contrast of a digital image. Guided lightwave transmission via frequency selective waveguides involves the emerging field of fiber optics and the ability of certain glassy compositions as a transmission medium for a range of frequencies simultaneously (multi-mode optical fiber) with little or no interference between competing wavelengths or frequencies. This resonant mode of energy and data transmission via electromagnetic (light) wave propagation, though low powered, is virtually lossless. Optical waveguides are used as components in Integrated optical circuits (e.g. light-emitting diodes, LEDs) or as the transmission medium in local and long haul optical communication systems. Also of value to the emerging materials scientist is the sensitivity of materials to radiation in the thermal infrared (IR) portion of the electromagnetic spectrum. This heat-seeking ability is responsible for such diverse optical phenomena as Night-vision and IR luminescence.

Thus, there is an increasing need in the military sector for high-strength, robust materials which have the capability to transmit light (electromagnetic waves) in the visible (0.4 – 0.7 micrometers) and mid-infrared (1 – 5 micrometers) regions of the spectrum. These materials are needed for applications requiring transparent armor, including next-generation high-speed missiles and pods, as well as protection against improvised explosive devices (IED).

In the 1960s, scientists at General Electric (GE) discovered that under the right manufacturing conditions, some ceramics, especially aluminium oxide (alumina), could be made translucent. These translucent materials were transparent enough to be used for containing the electrical plasma generated in high-pressure sodium street lamps. During the past two decades, additional types of transparent ceramics have been developed for applications such as nose cones for heat-seeking missiles, windows for fighter aircraft, and scintillation counters for computed tomography scanners.

In the early 1970s, Thomas Soules pioneered computer modeling of light transmission through translucent ceramic alumina. His model showed that microscopic pores in ceramic, mainly trapped at the junctions of microcrystalline grains, caused light to scatter and prevented true transparency. The volume fraction of these microscopic pores had to be less than 1% for high-quality optical transmission.

This is basically a particle size effect. Opacity results from the incoherent scattering of light at surfaces and interfaces. In addition to pores, most of the interfaces in a typical metal or ceramic object are in the form of grain boundaries which separate tiny regions of crystalline order. When the size of the scattering center (or grain boundary) is reduced below the size of the wavelength of the light being scattered, the scattering no longer occurs to any significant extent.

In the formation of polycrystalline materials (metals and ceramics) the size of the crystalline grains is determined largely by the size of the crystalline particles present in the raw material during formation (or pressing) of the object. Moreover, the size of the grain boundaries scales directly with particle size. Thus a reduction of the original particle size below the wavelength of visible light (~ 0.5 micrometers for shortwave violet) eliminates any light scattering, resulting in a transparent material.

Recently, Japanese scientists have developed techniques to produce ceramic parts that rival the transparency of traditional crystals (grown from a single seed) and exceed the fracture toughness of a single crystal. In particular, scientists at the Japanese firm Konoshima Ltd., a producer of ceramic construction materials and industrial chemicals, have been looking for markets for their transparent ceramics.

Livermore researchers realized that these ceramics might greatly benefit high-powered lasers used in the National Ignition Facility (NIF) Programs Directorate. In particular, a Livermore research team began to acquire advanced transparent ceramics from Konoshima to determine if they could meet the optical requirements needed for Livermore's Solid-State Heat Capacity Laser (SSHCL). Livermore researchers have also been testing applications of these materials for applications such as advanced drivers for laser-driven fusion power plants.

Examples

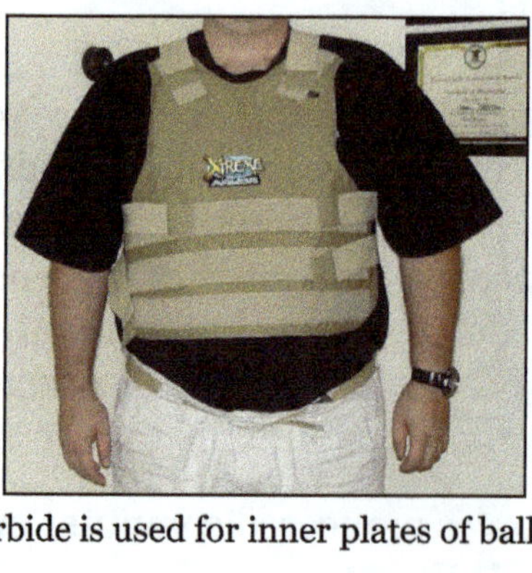

Silicon carbide is used for inner plates of ballistic vests.

Porcelain high-voltage insulator.

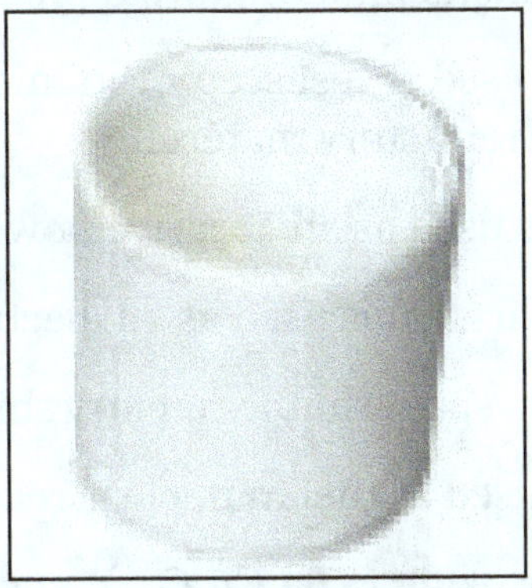

Ceramic BN crucible.

Until the 1950s, the most important ceramic materials were (1) pottery, bricks and tiles, (2) cements and (3) glass. A composite material of ceramic and metal is known as cermet.

Other ceramic materials, generally requiring greater purity in their make-up than those above, include forms of several chemical compounds, including:

- Barium titanate (often mixed with strontium titanate) displays ferroelectricity, meaning that its mechanical, electrical, and thermal responses are coupled to one another and also history-dependent. It is widely used in electromechanical transducers, ceramic capacitors, and data storage elements. Grain boundary conditions can create PTC effects in heating elements.
- Bismuth strontium calcium copper oxide, a high-temperature superconductor
- Boron oxide is used in body armor.
- Boron nitride is structurally isoelectronic to carbon and takes on similar physical forms: a graphite-like one used as a lubricant, and a diamond-like one used as an abrasive.

- Earthenware used for domestic ware such as plates and mugs.
- Ferrite is used in the magnetic cores of electrical transformers and magnetic core memory.
- Lead zirconate titanate (PZT) was developed at the United States National Bureau of Standards in 1954. PZT is used as an ultrasonic transducer, as its piezoelectric properties greatly exceed those of Rochelle salt.
- Magnesium diboride (MgB_2) is an unconventional superconductor.
- Porcelain is used for a wide range of household and industrial products.
- Sialon (Silicon Aluminium Oxynitride) has high strength; resistance to thermal shock, chemical and wear resistance, and low density. These ceramics are used in non-ferrous molten metal handling, weld pins and the chemical industry.
- Silicon carbide (SiC) is used as a susceptor in microwave furnaces, a commonly used abrasive, and as a refractory material.
- Silicon nitride (Si_3N_4) is used as an abrasive powder.
- Steatite (magnesium silicates) is used as an electrical insulator.
- Titanium carbide Used in space shuttle re-entry shields and scratchproof watches.
- Uranium oxide (UO_2), used as fuel in nuclear reactors.
- Yttrium barium copper oxide ($YBa_2Cu_3O_{7-x}$), another high temperature superconductor.
- Zinc oxide (ZnO), which is a semiconductor, and used in the construction of varistors.
- Zirconium dioxide (zirconia), which in pure form undergoes many phase changes between room temperature and practical sintering temperatures, can be chemically "stabilized" in several different forms. Its high oxygen ion conductivity recommends it for use in fuel cells and automotive oxygen sensors. In another variant, metastable structures can impart transformation toughening for mechanical applications; most ceramic knife blades are made of this material.
- Partially stabilised zirconia (PSZ) is much less brittle than other ceramics and is used for metal forming tools, valves and liners, abrasive slurries, kitchen knives and bearings subject to severe abrasion.

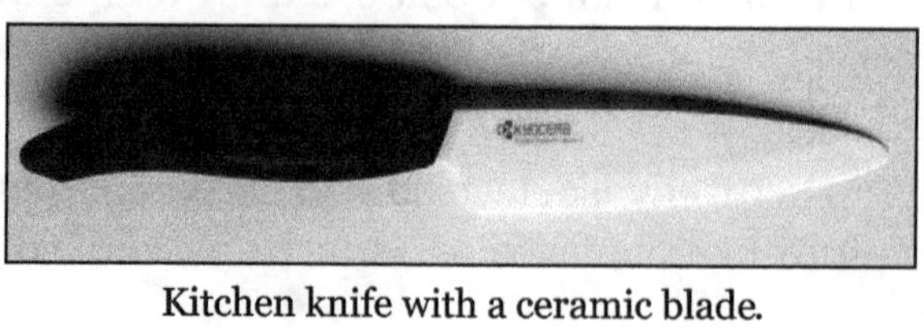

Kitchen knife with a ceramic blade.

Ceramic Products

By Usage

For convenience, ceramic products are usually divided into four main types; these are shown below with some examples:

- Structural, including bricks, pipes, floor and roof tiles.
- Refractories, such as kiln linings, gas fire radiants, steel and glass making crucibles.
- Whitewares, including tableware, cookware, wall tiles, pottery products and sanitary ware.
- Technical, also known as engineering, advanced, special, and fine ceramics. Such items include:
    - gas burner nozzles
    - ballistic protection
    - nuclear fuel uranium oxide pellets
    - biomedical implants
    - coatings of jet engine turbine blades
    - ceramic disk brake
    - missile nose cones
    - bearing (mechanical)
    - tiles used in the Space Shuttle program

Ceramics Made with Clay

Frequently, the raw materials of modern ceramics do not include clays. Those that do are classified as follows:

- Earthenware, fired at lower temperatures than other types
- Stoneware, vitreous or semi-vitreous
- Porcelain, which contains a high content of kaolin
- Bone china

Classification of Technical Ceramics

Technical ceramics can also be classified into three distinct material categories:

- Oxides: alumina, beryllia, ceria, zirconia
- Nonoxides: carbide, boride, nitride, silicide
- Composite materials: particulate reinforced, fiber reinforced, combinations of oxides and nonoxides.

Each one of these classes can develop unique material properties because ceramics tend to be crystalline.

Applications

- Knife blades: the blade of a ceramic knife will stay sharp for much longer than that of a steel knife, although it is more brittle and can snap from a fall onto a hard surface.
- Carbon-ceramic brake disks for vehicles are resistant to brake fade at high temperatures.
- Advanced composite ceramic and metal matrices have been designed for most modern armoured fighting vehicles because they offer superior penetrating resistance against shaped charges (such as HEAT rounds) and kinetic energy penetrators.
- Ceramics such as alumina and boron carbide have been used in ballistic armored vests to repel large-caliber rifle fire. Such plates are known commonly as small arms protective inserts, or SAPIs. Similar material is used to protect the cockpits of some military airplanes, because of the low weight of the material.
- Ceramics can be used in place of steel for ball bearings. Their higher hardness means they are much less susceptible to wear and typically last for triple the lifetime of a steel part. They also deform less under load, meaning they have less contact with the bearing retainer walls and can roll faster. In very high speed applications, heat from friction during rolling can cause problems for metal bearings, which are reduced by the use of ceramics. Ceramics are also more chemically resistant and can be used in wet environments where steel bearings would rust. In some cases, their electricity-insulating properties may also be valuable in bearings. Two drawbacks to ceramic bearings are a significantly higher cost and susceptibility to damage under shock loads.
- In the early 1980s, Toyota researched production of an adiabatic engine using ceramic components in the hot gas area. The ceramics would have allowed temperatures of over 3000 °F (1650 °C). The expected advantages would have been lighter materials and a smaller cooling system (or no need for one at all), leading to a major weight reduction. The expected increase of fuel efficiency of the engine (caused by the higher temperature, as shown by Carnot's theorem) could not be

verified experimentally; it was found that the heat transfer on the hot ceramic cylinder walls was higher than the transfer to a cooler metal wall as the cooler gas film on the metal surface works as a thermal insulator. Thus, despite all of these desirable properties, such engines have not succeeded in production because of costs for the ceramic components and the limited advantages. (Small imperfections in the ceramic material with its low fracture toughness lead to cracks, which can lead to potentially dangerous equipment failure.) Such engines are possible in laboratory settings, but mass production is not feasible with current technology.

- Work is being done in developing ceramic parts for gas turbine engines. Currently, even blades made of advanced metal alloys used in the engines' hot section require cooling and careful limiting of operating temperatures. Turbine engines made with ceramics could operate more efficiently, giving aircraft greater range and payload for a set amount of fuel.
- Recent advances have been made in ceramics which include bioceramics, such as dental implants and synthetic bones. Hydroxyapatite, the natural mineral component of bone, has been made synthetically from a number of biological and chemical sources and can be formed into ceramic materials. Orthopedic implants coated with these materials bond readily to bone and other tissues in the body without rejection or inflammatory reactions so are of great interest for gene delivery and tissue engineering scaffolds. Most hydroxyapatite ceramics are very porous and lack mechanical strength, and are used to coat metal orthopedic devices to aid in forming a bond to bone or as bone fillers. They are also used as fillers for orthopedic plastic screws to aid in reducing the inflammation and increase absorption of these plastic materials. Work is being done to make strong, fully dense nanocrystalline hydroxyapatite ceramic materials for orthopedic weight bearing devices, replacing foreign metal and plastic orthopedic materials with a synthetic, but naturally occurring, bone mineral. Ultimately, these ceramic materials may be used as bone replacements or with the incorporation of protein collagens, synthetic bones.
- High-tech ceramic is used in watchmaking for producing watch cases. The material is valued by watchmakers for its light weight, scratch resistance, durability and smooth touch. IWC is one of the brands that initiated the use of ceramic in watchmaking.

Ceramics in Archaeology

Ceramic artifacts have an important role in archaeology for understanding the culture, technology and behavior of peoples of the past. They are among the most common artifacts to be found at an archaeological site, generally in the form of small fragments of broken pottery called sherds. Processing of collected sherds can be consistent with two main types of analysis: technical and traditional.

Traditional analysis involves sorting ceramic artifacts, sherds and larger fragments into specific types based on style, composition, manufacturing and morphology. By creating these typologies it is possible to distinguish between different cultural styles, the purpose of the ceramic and technological state of the people among other conclusions. In addition, by looking at stylistic changes of ceramics over time is it possible to separate (seriate) the ceramics into distinct diagnostic groups (assemblages). A comparison of ceramic artifacts with known dated assemblages allows for a chronological assignment of these pieces.

The technical approach to ceramic analysis involves a finer examination of the composition of ceramic artifacts and sherds to determine the source of the material and through this the possible manufacturing site. Key criteria are the composition of the clay and the temper used in the manufacture of the article under study: temper is a material added to the clay during the initial production stage, and it is used to aid the subsequent drying process. Types of temper include shell pieces, granite fragments and ground sherd pieces called 'grog'. Temper is usually identified by microscopic examination of the temper material. Clay identification is determined by a process of refiring the ceramic, and assigning a color to it using Munsell Soil Color notation. By estimating both the clay and temper compositions, and locating a region where both are known to occur, an assignment of the material source can be made. From the source assignment of the artifact further investigations can be made into the site of manufacture.

Dimension Stone

Large blocks of granite dimension stone being loaded at Teignmouth in 1827.

Dimension stone is natural stone or rock that has been selected and finished (i.e., trimmed, cut, drilled, ground, or other) to specific sizes or shapes. Color, texture and pattern, and surface finish of the stone are also normal requirements. Another important selection criterion is durability: the time measure of the ability of dimension stone to endure and to maintain its essential and distinctive characteristics of strength, resistance to decay, and appearance.

Quarries that produce dimension stone or crushed stone (used as construction aggregate) are interconvertible. Since most quarries can produce either one, a crushed stone quarry can be converted to dimension stone production. However, first the stone shattered by heavy and indiscriminate blasting must be removed. Dimension stone is separated by more precise and delicate techniques, such as diamond wire saws, diamond belt saws, burners (jet-piercers), or light and selective blasting with Primacord, a weak explosive.

Stone and Rock Types

A variety of igneous, metamorphic, and sedimentary rocks are used as structural and decorative dimension stone. These rock types are more commonly known as granite, limestone, marble, travertine, quartz-based stone (sandstone, quartzite) and slate. Other varieties of dimension stone that are normally considered to be special minor types include alabaster (massive gypsum), soapstone (massive talc), serpentine and various products fashioned from natural stone.

A variety of finishes can be applied to dimension stone to achieve diverse architectural and aesthetic effects. These finishes include, but are not limited to, the following. A polished finish gives the surface a high luster and strong reflection of incident light (almost mirror-like). A honed finish provides a smooth, satin-like ("eggshell"), non-reflective surface. More textured finishes include brush-hammered, sandblasted, and thermal. A brush-hammered finish, similar to a houndstooth pattern, creates a rough, but uniformly patterned surface with impact tools varying in coarseness. A sandblasted surface provides an irregular pitted surface by impacting sand or metal particles at high velocity against a stone surface. A thermal (or flamed) finish produces a textured, nonreflective surface with only a few reflections from cleavage faces, by applying a high-temperature flame. This finish may change the natural color of the stone depending on mineralogical composition, particularly with stones containing higher levels of iron.

The most easily accessible general (non-graphic) references are the latest Minerals Yearbook Chapter (production and foreign trade, with statistics), and the latest (Issue 31) Dimension Stone Advocate News (new "building green" developments and demand statistics). The most comprehensive, graphic references are Natural Stone Database by Abraxas Verlag (www.natural-stone-database.com), "Dimension Stones of the World, Volumes I & II" (Marble Institute of America) and "Natural Stones Worldwide CD" .

Major Applications

While common colors used in some of the major applications are listed below, there is an extraordinarily wide range of colours, available in thousands of patterns. These patterns are created by geological phenomena such as mineral grains, inclusions, veins, cavity fillings, blebs, and streaks. In addition, rocks and stones not normally classed as dimension stone are sometimes selected for these applications. These can included tiles made of jade, agate, and jasper.

Rough cut slabs of granite dimension stone.

Stone (usually granite) countertops and bathroom vanities both involve a finished slab of stone, usually polished but sometimes with another finish (such as honed or sand-blasted). Industry standard thicknesses in the United States are 3/4" (2 cm) and 1.25" (3 cm). Often 2 cm slabs will be laminated at the edge to create the appearance of a thicker edge profile. The slabs are cut to fit the top of the kitchen or bathroom cabinet, by measuring, templating or digital templating. Countertop slabs are commonly sawn from rough blocks of stone by reciprocating gangsaws using steel shot as abrasive. More modern technology utilizes diamond wire saws which use less water and energy. Multi-wire saws with as many as 60 wires can slab a block in less than two hours. The slabs are finished (i.e., polished, honed), then sealed with resin to fill micro-fissures and surface imperfections typically due to the loss of poorly bonded elements such as biotite. The fabricators shop cuts these slabs down to final size and finishes the edges with equipment such as hand-held routers, grinders, CNC equipment, or polishers. In 2008, concerns were raised regarding radon emissions from granite countertops; the National Safety Council states that the contributions of radon to inside air come from the soil and rock around the residence (69%), the outdoor air and the water supply (28%), and only 2.5% from all building materials-including granite countertops. A concerned homeowner can employ ASTM radon mitigation and removal techniques. The stone for countertops or vanities is usually granite, but often is marble (especially for vanity tops), and is sometimes limestone or slate. The majority of the stone for this application is produced in Brazil, Italy, and China.

Tile is a thin modular stone unit, commonly 12 in. square (30.5 cm) and 3/8 in. (10 mm) deep. Other popular sizes are 15 in. square (38 cm), 18 in. square (46 cm), and 24 in. square (61 cm); these will usually be deeper than the 12 in. square. The majority of tile has a polished finish, but other finishes such as honed are becoming more common. Almost all stone tile is mass-produced by automated tile lines to identical size, finish, and close tolerances. Exceptions include slate flooring tile and special orders: tile with odd sizes or shapes, unusual finishes, or inlay work. In summary, the automated tile line is a complicated complex of cutting and calibrating machines, honing-polishing machines, edging machines that put on flat or rounded edges, and interconnecting conveyors to move the stone from the slab input to the final tile product. The stone for tiles is most commonly marble, but often is granite, and sometimes limestone, slate, or quartz-based stone. Common colors are white and light earth colors. Much of the stone for this application is produced in Italy and China.

Slate tile covers this entire structure in Germany.

Stone monuments include tombstones, grave markers or as mausoleums. After being gangsawed into big deep (up to 10 ft (3.0 m) wide and over 6 inches deep) slabs, smaller saws or guillotines (they break the granite and make the rough edges commonly seen on monuments) shape the monuments. The fronts and backs are usually polished. The individual monuments are then carved, shaped, and further defined by hand tools and sandblasting equipment. At this time, the stone for monuments is most commonly granite, sometimes marble (as in military cemeteries), and rarely others. Granite and quartz both demonstrate good durability, especially because rain is naturally acidic. (This is a natural consequence of the carbon dioxide present in the atmosphere, which generates a weak solution of carbonic acid in rainfall; further acidification of rainfall arises from oxides of sulphur and nitrogen due to anthropogenic emissions). Long-term durability of quartz monuments can be observed at the Harpers Ferry cemetery, where an old rose quartz tombstone stands. (Limestone and sandstone were commonly chosen for monuments in the nineteenth century, but they are no widely longer used because of the rapid erosion rates due to dissolution of acid-vulnerable carbonates by acidic rainfall.) The most common monument colors for granite are gray, black, and mahogany; for marble, white is most popular. Today, the majority of the stone used in

North America in this application is imported from countries such as India and China. This has depressed traditional North American monument centers such as Georgia and Quebec.

Dimension stone has been used in the construction of buildings for centuries. Due to costs, today stone veneers are usually used in place of solid stone blocks. This courthouse was built of dimension stone quarried in Berea, Ohio.

There are a number of smaller applications for buildings and traffic-related uses. Building components include stone used as veneer (exterior), ashlar, or other shapes. Veneer is a non load-bearing facing of stone attached to a backing of an ornamental nature, although it also protects and insulates. Ashlar is a square block of stone, often brick-sized, for facing of walls (primarily exterior). The other shapes are rectangular blocks used for stair treads, sills, and coping (coping is sometimes nonrectangular). The shapes subject to foot traffic will usually have an abrasive finish such as honed or sandblasted. The stone is mostly limestone, but often is quartz-based stone (sandstone), or even marble or granite. Roofing slate is a thin-split shingle-sized piece of slate, and when in place forms the most permanent kind of roof; slate is also used as countertops and flooring tile. Traffic-related stone is that which is used for curbing (vehicular) and flagstone (pedestrian). Curbing is thin stone slabs used along streets or highways to maintain the integrity of sidewalks and borders. Flagstone is a shallow naturally irregular-edged slab of stone, sometimes sawed into a rectangular shape, used as paving (almost always pedestrian). For curbing, the stone is almost always granite, and for flagstone the stone is almost always quartz-based stone (sandstone or quartzite).

There are several other applications resembling flagstone in using rough dimension (or crushed) stone, usually as quarried, sometimes made smaller (i.e. by a jackhammer), often simply put in place: dry stone and riprap.

The stone used in these applications usually has to have certain properties, or meet a standard specification. The American Society for Testing and Materials (ASTM) has such specifications for granite, marble, limestone, quartz-based dimension stone (C616), slate (C629), travertine (C1527), and serpentine (C1526).

Production

Marble quarry in Carrara, Italy.

The major producers of dimension stone include Brazil, China, India, Italy, and Spain, and each have annual production levels of nine to over twenty-two million tons. Portugal produces 3 million tons of dimension stone each year.

According to the USGS, 2007 U.S. dimension stone production was 1.39 million tons valued at $275 million, compared to 1.33 million tons (revised) valued at $265 million in 2006. Of these, granite production was 453,000 tons valued at $106 million in 2007 and 428,000 tons valued at $105 million in 2006, and limestone was 493,000 tons valued at $93.3 million in 2007 and 559,000 tons valued at $96.1 million in 2006. The United States is at best a mid-level dimension stone producer on the world scene; Portugal produces twice as much dimension stone annually.

World comparison for dimension stone demand: The DSAN World Demand Index for (finished) Granite was 227 in 2006, 247 in 2007, and 249 in 2008, and the World Demand Index for (finished) Marble was 200 in 2006, 248 in 2007, and 272 in 2008. The DSAN World Demand for (finished) Granite Index showed a growth of 12% annually for the 2000-2008 period, compared to 14% annually for the 2000-2007 period, and compared to 15% annually for the 2000-2006 period. The DSAN World Demand for (finished) Marble Index showed a growth of 13.5% annually for the 2000-2008 period, compared to 14.0% annually for the 2000-2007 period, and compared to 12.5% annually for the 2000-2006 period. The indexes show world demand for granite has clearly been weakening since 2006, while the world demand for marble only weakened from 2007 to 2008. Other DSAN indexes for 2008 indicate that the 2000-2008 growth was down from the 2000-2007 growth.

The DSAN U.S. Ceramic Tile Demand Index shows a drop of 4.8% annually for the 2000-2007 period, compared to growth of 5.0% annually for the 2000-2006 period. The "traditional" major ceramic tile suppliers, Italy and Spain, have been losing markets to new entrants Brazil and China. The same thing has been happening with dimension stone with increasing supplies from Brazil, China and India.

In 2008, Chinese exports of granite countertops and marble tile increased from 2007, while those of Italy and Spain did not In early 2009, the Chinese Government has a hands-off policy towards its dimension stone industry.

"Building green" with Dimension Stone

Marble cladding on a building.

Green building or environmentally friendly construction with natural materials, is an idea that has been around for several decades. Energy price increases and the need for energy conservation when heating or cooling buildings have recently brought it to the fore. This resulted in the formation in 1993 of the U.S. Green Building Council (USGBC), which has developed a building rating system called Leadership in Energy and Environmental Design (LEED). Educational institutions (colleges, universities, grade, and high schools) are often requiring new buildings to be green, and a few jurisdictions (i.e., some cities) have some rules promoting green building. When "building green", dimension stone has a big advantage over steel, concrete, glazed glass and laminated plastics, whose productions are all energy intensive and create significant air and water pollution. As an entirely natural product, dimension stone also has an advantage over synthetic/artificial stone products, as well as composite and space-age materials. One LEED requirement provides that the dimension stone used in a green building be quarried within a 500-mile (800 km) radius of the building being constructed. This gives a clear advantage to domestic dimension stone, plus some quarried near the U.S. borders with Canada and Mexico. A current problem is how to consider stone quarried domestically, sent to China or Italy for finishing, and shipped back to be used in a project.

When demolishing a structure, dimension stone is 100% reusable and can be salvaged for new construction, used as paving or crushed for use as aggregates. There are also "green" methods of stone cleaning either in development or already in use, such as removing the black gypsum crusts that form on marble and limestone by applying sulfate-reducing bacteria to the crust to gasify it, breaking up the crust for easy removal.

The Federal Trade Commission (FTC) in America is re-examining and will most likely update its "Green Guides" used to regulate green advertising claims. The updating will emphasize green building, including the products it involves, such as dimension stone. When the new requirements are finalized, the FTC will go after firms that violate the new requirements, in order to establish legal precedents.

The Natural Stone Council has a library of information on building green with dimension stone, including life-cycle inventory data for each major dimension stone, giving the amount of energy, water, other inputs, and processing emissions, plus some best practice studies. In addition, it has shown ways that dimension stone can contribute LEED points, such as using a light-colored dimension stone to reduce heat-island effects, using dimension stone's thermal mass to impact indoor ambient air temperature thereby increasing energy efficiency, and especially by reusing dimension stone rather sending it to the landfill.

Sustainability of Dimension Stone

Dimension stone is one of the most sustainable of the industrial minerals since it is created by separating it from the natural bedrock underlying all land on every continent. Dimension stone rates very well in terms of the criteria on the ASTM checklist for sustainability of building products: there are no toxic materials used in its processing, there are no direct greenhouse gas emissions during processing, the dust created is controlled, the water used is almost completely recycled (per OSHA/MSHA regulation), and it is a perpetual resource (virtually inexhaustible in a human time scale). Dimension stone in use can last many generations, even centuries, so the dimension stone manufacturers have not needed a product recycling program. However, there are practical qualifications to and constraints on that sustainability. The dimension stone color and pattern can be changed by weathering when it is very near the surface. The color and pattern can also be changed by proximity to an igneous rock body or by the presence of circulating groundwater charged with carbon dioxide (i.e., limestone, travertine, marble). On the other hand, changes in color and/or pattern can be positive. For example, there are at least 14 separately trade-named varieties of Carrara Marble with many patterns (or no patterns) ranging in shade from white to gray. The presence of faults or closely spaced joints can render the stone unusable. These faults and joints do not have to be at odd angles in the stone mass. Closely spaced, wrongly spaced, or nonparallel bedding planes can make the stone unusable, particularly if the bedding planes are planes of weakness. If part of the stone in one area is unusable, there will be another usable part of the stone elsewhere in the formation. A quarry is not a short-term project unless it encounters one of these constraints. Examples of big, old quarries operating for more than a century include the Barre (VT) granite quarry, the Georgia Marble quarry at Tate, several of the Carrara (Italy) marble quarries, and the Penrhyn (Wales) slate quarry. A quarry will produce dust, noise, and some water pollution, but these can be remedied without too much trouble. The landscape may also have to be restored if quarry waste is temporarily or permanently placed on adjacent land.

Stone Recycling and Reuse

Reconstruction of the Charles Bridge in Prague showing numbered dimension blocks.

Recycling dimension stone can occur when structures are demolished, along with recycling timber and recycling construction aggregate in the form of concrete. The material most likely to be recycled is concrete, and this represents the largest volume of recycled construction material. Not too many structures incorporate dimension stone, and even fewer of them have dimension stone worth saving. Stone recycling is usually done by specialists that monitor local demolition activity, looking for stone-containing houses, buildings, bridge abutments, and other dimension stone structures scheduled for demolition. Particularly treasured are old hand-carved stone pieces with the chisel marks still on them, local stones no longer quarried or that are quarried in a different shade of color or appearance. There is no national or regional trade in reclaimed stone, so a large storage yard is required, since the recovered stone may not be quickly sold and reused. The recycled dimension stone is used in old stone buildings being renovated (to replace deteriorated stone pieces), in fireplace mantels, benches, veneer, or for landscaping (like for retaining walls).

The Parthenon in Athens underwent a major reconstruction prior to the 2004 Olympics.

Related to stone recycling and stone reuse is the deconstruction and reconstruction of a stone building. The building is taken apart stone block by stone block and the location and orientation of each block is carefully noted. Any roofing slate and interior stone in place is catalogued and moved in the same fashion. After the blocks, slate, and other stone used have been transported to the new location, they are put back in place where and how they were originally, thus reassembling the building. This is typically very expensive and rare but valuable in terms of historic preservation.

Dimension stone is also reused. Buildings immediately spring to mind, but such things as the ornate stone walls, arches, stairways and balustrades alongside a boulevard can also be renovated and reused. Sometimes the old interior of the building is kept as is, after repair. Sometimes the old building is gutted, leaving only a shell or facade and the space inside reconfigured and modernized. The stone work will usually need attention too.

The old stone work may only need cleaning or sandblasting, but it may need more. Firstly, the building exterior (facade) needs to be inspected for unsafe conditions. Next, the building walls need to be inspected for water leakages. The most likely needs are mortar restoration (repointing), applying consolidants to the old stone, or replacing pieces of stone that are deteriorated (damaged) beyond the point of any repair. The repointing is the removal of existing damaged mortar from the outer portion of the joint between stone units and its replacement by new mortar matching the appearance of the old. The consolidants re-establish the original natural bonding between the stone particles that weathering has removed. Deteriorated pieces of stone work are replaced with pieces of stone that match the original as much as possible. Exterior dimension stone will often change color after exposure to weather over time. For example, Indiana Limestone will weather from a tan to an attractive light yellow. Interior dimension stone can sometimes change its shade a little over time too. For both, it may not be possible to find an exact match, even from the original quarry. Stone will often change its appearance from location to location in the same quarry. If the dimension stone renovationist is truly fortunate, the original builder put aside some spare pieces of the stone for future need.

Stone: Life-cycle Assessment and Best Practices

As in every economic sector, the construction industry's purchases of materials and services creates a whole chain of processes from raw material selecting in situ, removal from the earth, usually proceeding to cutting, finishing, or processing/manufacturing, then transport, and retailing. All of these activities have significant upstream (off-site) environmental impacts, whether in terms of energy and raw resource use or emissions to air, land, or water impacting living organisms or the Earth's surface (non-organic). Life cycle assessment is a method for estimating and comparing a range of environmental performance measures (e.g. global warming, acidification potential, toxicity, ozone depletion potentials) over the full life cycle of a product, a building assembly, or a whole building. As such, it provides a comprehensive means for evaluating and comparing products rather than prescriptive measures of individual product characteristics.

The ASTM has some relevant standards, particularly a guide on environmental life cycle assessment of building materials/products (E1991) that shows how to minimize the subjectivity that commonly mars and confuses environmental decision making. In particular, this guide describes the inventory analysis phase that requires data that is suitable for its intended purpose, thus covering data quality (such as completeness, reliability, accuracy, and credibility) as well as the allocation of the data (for multiple inputs and outputs), among other things. Results have to be on a common basis to allow a statistically significant comparison of alternative building product differences in the interpretation.

The Natural Stone Council (NSC) has commissioned some life-cycle inventory data for use in life cycle assessments. Almost 90% of the effort in doing a life cycle assessment involves getting reliable data. For example, the NSC has data that the Global Warming Potential for granite quarrying is 100 kg of carbon dioxide equivalents and for granite processing is 500 (same units); and the Global Warming Potential for limestone quarrying is 20 kg carbon dioxide equivalents while for limestone processing it is 80 (same units). The data on energy and water use include everything back to removal of overburden in the dimension stone quarry and upstream production of energy and fuels, and forward to packaging of finished dimension stone product or slabs for shipment and transport, or to moving scrap stone to storage or reclamation and to capturing and treatment of dust and waste water. The data is then placed in an impact category (i.e. changes to air, changes to water), characterized as to the contribution of the item to the impact compared to other items, and then the impact categories are assigned weights among themselves to show their relative importance.

The Natural Stone Council has also commissioned four Best Practices. One is on water consumption, treatment, and reuse while extracting and processing dimension stone, including dust mitigation, sludge management, and maximizing water recycling. Another is on site maintenance and quarry closure, including minimizing dust, noise, vibration and keeping the operation clean and tidy, both of which help in restoring the surface upon quarry closure. A third one is on solid waste management, including overburden, damaged stone unsaleable as product, sludge deposited from waste water, spent or spilled petroleum products, or metal scrap. The fourth one is on efficiently transporting stone to be finished as products, then transporting the products to consumers by centralizing freight management, consolidating small loads, choosing appropriate trucks, balancing and securing the load, and packaging with sustainable materials.

Stone Selection and Cleaning

The selector of dimension stone begins by considering stone color and appearance, and how the stone will match its surroundings. The selector has literally thousands of options to choose from, and should examine many options. In addition to many hundreds of different stones with different colors and patterns, each stone can change radically in color and appearance when a different finish is put on it. A polished finish accentuates the color and makes any pattern more vivid, and the rougher finishes (i.e. honed, thermal) lighten the color and make the patterns more subdued. With thousands of possibilities, the selector must start by looking at many stones in many different finishes, or photos of them. Such photos can be found on some dimension stone websites, and on DSAN's Architects Stone Selection Helper.

In addition to selecting a stone color and pattern, the suitability of its properties for the intended use must be considered. Stone being chosen for countertops or vanities should be nonabsorptive, resist stains, and be heat and impact resistant. Stone being used in tiles should be sealed in order to resist staining by spilled liquids. Stone being used for flooring, paving, or surfaces subject to foot or vehicular traffic ought to have a semiabrasive finish for slip resistance, such as bush-hammered or thermal. A glossy polished finish will be slick. Most flagstone surfaces are rough enough to be naturally slip-resistant.

Dimension stone requires some specialized methods for cleaning and maintenance. Abrasive cleaners should not be used on a polished stone finish because it will wear the polish off. Acidic cleaners can not be used on marble or limestone because it will remove (i.e. dissolve) the finish. Textured finishes (thermal, bush-hammered) can be treated with some mildly abrasive cleaners but not bleach or an acidic cleaner (if marble or limestone). Stains are another consideration; stains can be organic (food, grease, or oil) or metallic (iron, copper). Stains require some special removal techniques, such as the poultice method. A new method of cleaning stone on ancient buildings (medieval and renaissance) has been developed in Europe: sulfur-reducing bacteria are used on the black gypsum-containing crusts that form on such buildings to convert the sulfur to a gas that dissipates, thus destroying the crust while leaving the patina produced by aging on the underlying stone. This method is still in development and not yet commercially available.

Stone Finishes

The surface of a stone may be finished in a variety of ways. Below are some typical terms:

- Polished finish - a glossy surface which brings out the full color and character of the stone.
- Hone finish - a satin smooth surface finish with little or no gloss. This is recommended for commercial floors.
- Thermal finish - a surface treatment applied by intense heat flaming.
- Diamond sawed - finish produced by sawing with a diamond toothed saw.

- Rough sawn - a surface finish resulting from the gang sawing (or frame saw) process.
- Brush-hammered - a mechanical process which produces textured surfaces. Texture varies from subtle to rough.

References

- ASTM, E06, "E2121-08 Standard Practice for Installing Radon Mitigation Systems in Existing Low-Rise Residential Buildings", ASTM, 2008, pp. 644-656 ISBN 978-0-8031-5768-2
- Carter, C. B.; Norton, M. G. (2007). Ceramic materials: Science and engineering. Springer. pp. 3 & 4. ISBN 978-0-387-46271-4.
- L. Mead and G.S. Austin "Dimension Stone", Industrial Minerals and Rocks, 7th Edition, Littleton CO: AIME-Society of Mining Engineers, 2005, pp. 907-923 ISBN 0-87335-233-5
- Black, J. T.; Kohser, R. A. (2012). DeGarmo's materials and processes in manufacturing. Wiley. p. 226. ISBN 978-0-470-92467-9.
- ASTM, C18, "C1496-01 Standard Guide for Assessment and Maintenance of Exterior Dimension Stone Masonry Walls and Facades", ASTM, 2007, pp. 519-523 ISBN 0-8031-4104-1
- Carter, C. B.; Norton, M. G. (2007). Ceramic materials: Science and engineering. Springer. pp. 20 & 21. ISBN 978-0-387-46271-4.

Permissions

All chapters in this book are published with permission under the Creative Commons Attribution Share Alike License or equivalent. Every chapter published in this book has been scrutinized by our experts. Their significance has been extensively debated. The topics covered herein carry significant information for a comprehensive understanding. They may even be implemented as practical applications or may be referred to as a beginning point for further studies.

We would like to thank the editorial team for lending their expertise to make the book truly unique. They have played a crucial role in the development of this book. Without their invaluable contributions this book wouldn't have been possible. They have made vital efforts to compile up to date information on the varied aspects of this subject to make this book a valuable addition to the collection of many professionals and students.

This book was conceptualized with the vision of imparting up-to-date and integrated information in this field. To ensure the same, a matchless editorial board was set up. Every individual on the board went through rigorous rounds of assessment to prove their worth. After which they invested a large part of their time researching and compiling the most relevant data for our readers.

The editorial board has been involved in producing this book since its inception. They have spent rigorous hours researching and exploring the diverse topics which have resulted in the successful publishing of this book. They have passed on their knowledge of decades through this book. To expedite this challenging task, the publisher supported the team at every step. A small team of assistant editors was also appointed to further simplify the editing procedure and attain best results for the readers.

Apart from the editorial board, the designing team has also invested a significant amount of their time in understanding the subject and creating the most relevant covers. They scrutinized every image to scout for the most suitable representation of the subject and create an appropriate cover for the book.

The publishing team has been an ardent support to the editorial, designing and production team. Their endless efforts to recruit the best for this project, has resulted in the accomplishment of this book. They are a veteran in the field of academics and their pool of knowledge is as vast as their experience in printing. Their expertise and guidance has proved useful at every step. Their uncompromising quality standards have made this book an exceptional effort. Their encouragement from time to time has been an inspiration for everyone.

The publisher and the editorial board hope that this book will prove to be a valuable piece of knowledge for students, practitioners and scholars across the globe.

Index

www.ingramcontent.com/pod-product-compliance
Lightning Source LLC
LaVergne TN
LVHW080613210926
848485LV00003B/154